T0233703

BOOTSTRAP TECHNIQUES FOR SIGNAL PROCESSING

A signal processing practitioner often asks themselves, "How accurate is my parameter estimator?" There may be no answer to this question if an analytic analysis is too cumbersome and the measurements sample is too small. The statistical bootstrap, an elegant solution, re-uses the original data with a computer to re-estimate the parameters and infer their accuracy.

This book covers the foundations of the bootstrap, its properties, its strengths, and its limitations. The authors focus on bootstrap signal detection in Gaussian and non-Gaussian interference as well as bootstrap model selection. The theory presented in the book is supported by a number of useful practical examples written in MATLAB.

The book is aimed at graduate students and engineers, and includes applications to real-world problems in areas such as radar, sonar, biomedical engineering and automotive engineering.

Abdelhak Zoubir is Professor of Signal Processing at Darmstadt University of Technology, Germany. He held positions in industry and in academia in Germany and Australia. He has published over 180 technical papers in the field of statistical methods for signal processing. He has maintained his research interest in the bootstrap since the late 1980s. He also regularly gives courses and tutorials on the bootstrap and its application for engineers.

D. Robert Iskander received a Ph.D. degree in signal processing from Queensland University of Technology (QUT), Australia, and holds the position of a principal research fellow in the Centre for Health Research, QUT. He has published over 60 technical papers in the field of statistical signal processing and its application to optometry. He also has several patents in the area of visual optics.

BOOTSTRAP TECHNIQUES FOR SIGNAL PROCESSING

ABDELHAK M. ZOUBIR
D. ROBERT ISKANDER

CAMBRIDGE
UNIVERSITY PRESS

CAMBRIDGE
UNIVERSITY PRESS

University Printing House, Cambridge CB2 8BS, United Kingdom

One Liberty Plaza, 20th Floor, New York, NY 10006, USA

477 Williamstown Road, Port Melbourne, VIC 3207, Australia

314-321, 3rd Floor, Plot 3, Splendor Forum, Jasola District Centre, New Delhi - 110025, India

79 Anson Road, #06-04/06, Singapore 079906

Cambridge University Press is part of the University of Cambridge.

It furthers the University's mission by disseminating knowledge in the pursuit of
education, learning and research at the highest international levels of excellence.

www.cambridge.org
Information on this title: www.cambridge.org/9780521034050

© Cambridge University Press 2014

First published 2014

A catalogue record for this publication is available from the British Library

ISBN 978-0-521-83127-7 Hardback
ISBN 978-0-521-03405-0 Paperback

Contents

Contents

vii

Preface

The bootstrap genesis is generally attributed to Bradley Efron. In 1977 he wrote the famous Rietz Lecture on the estimation of sampling distributions based on observed data (Efron, 1979a). Since then, a number of outstanding and nowadays considered classical statistical texts have been written on the topic (Efron, 1982; Hall, 1992; Efron and Tibshirani, 1993; Shao and Tu, 1995), complemented by other interesting monographic exposés (LePage and Billard, 1992; Mammen, 1992; Davison and Hinkley, 1997; Manly, 1997; Barbe and Bertail, 1995; Chernick, 1999).

Efron and Tibshirani (1993) state in the Preface of their book *Our goal in this book is to arm scientists and* **engineers**, *as well as statisticians, with computational techniques that they can use to analyze and understand complicated data sets*. We share the view that Efron and Tibshirani (1993) have written an outstanding book which, unlike other texts on the bootstrap, is more accessible to an engineer. Many colleagues and graduate students of ours prefer to use this text as the major source of knowledge on the bootstrap. We believe, however, that the readership of (Efron and Tibshirani, 1993) is more likely to be researchers and (post-)graduate students in mathematical statistics than engineers.

To the best of our knowledge there are currently no books or monographs on the bootstrap written for electrical engineers, particularly for signal processing practitioners. Therefore the decision for us to fill such a gap. The bootstrap world is a great one and we feel strongly for its discovery by engineers. Our aim is to stimulate interest by engineers to discover the power of bootstrap methods. We chose the title *Bootstrap Techniques for Signal Processing* not only because we work in this discipline and because many of the applications in this book stem from signal processing problems, but also owing to the fact that signal processing researchers and (post-)graduate students are the more likely engineers to use the book. In particular, we would

like to reach researchers and students in statistical signal processing such as
those working on problems in areas that include radar, sonar, telecommunications and biomedical engineering.

We have made every attempt to convey the "how" and "when" to use
the bootstrap rather than mathematical details and proofs. The theory of
the bootstrap is well established and the texts mentioned above can give
the necessary details if the reader so wishes. We have included at least one
example for every introduced topic. Some of the examples are simple, such
as finding a confidence interval for the mean. Others are more complicated
like testing for zero the frequency response of a multiple-input single-output
linear time-invariant system.

It was difficult to decide whether we should include MATLAB† codes for
the examples provided. After some deliberation, and given the fact that
many graduate students and researchers ask for MATLAB codes to reproduce
published results, we decided to include them. We have also provided a
MATLAB toolbox which comprises frequently used routines. These routines
have been purposely written for the book to facilitate the implementation
of the examples and applications. All the MATLAB routines can be found in
the Appendices.

A few tutorial papers on the bootstrap for signal processing exist. The interested readers can refer to the work of Zoubir (1993); Zoubir and Boashash
(1998), and Zoubir (1999).

We are grateful to our colleagues Hwa-Tung Ong, Ramon Brcich, and
Christopher Brown for making very helpful comments and suggestions on
the manuscript. Additional words of thanks are for Hwa-Tung Ong for his
help in the development of the Bootstrap MATLAB Toolbox. We would like
to thank our current and past colleagues and graduate students who contributed directly or indirectly to the completion of the book. In particular,
we would like to thank Johann Böhme, Don Tufts, David Reid, Per Pelin,
Branko Ristic, Jonathon Ralston, Mark Morelande, Said Aouada, Amar
Abd El-Sallam and Luke Cirillo. The authors are grateful to all honours
and PhD students and colleagues of the Communications and Signal Processing Group at Curtin University of Technology in Perth, Australia and
special thanks are due to Tanya Vernon for her continued support to the
group.

Many government agencies and industries supported our research on the
bootstrap over the years. Thanks are due to Boualem Boashash at Queensland University of Technology and John Hullett and Zigmantas Budrikis at

---

† MATLAB is a registered trademark of The MathWorks, Inc.

Curtin University of Technology. The data from the JINDALEE over-the-horizon radar system was provided by the Defence Science and Technology Organisation (DSTO) in Edinburgh, South Australia. Special thanks are due to Stuart Anderson and Gordon Frazer for their generous support. The data from the GPR system was also collected at the DSTO in Edinburgh, South Australia. Words of gratitude are for Ian Chant and Canicious Abeynayake for their support with the landmine project. Zoubir acknowledges the financial support of the Australian Research Council. We also thank Michael Collins from the Centre for Eye Research at Queensland University of Technology for his support and encouragement during the preparation of this book. The encouragement of Phil Meyler from the Cambridge University Press is also gratefully acknowledged.

We thank our families, wives and children for their support, understanding and love. Without their patience this work could not be completed.

Last, we wish to refer the reader to a recent exposition by Bradley Efron (2002) on the role of bootstrap methods in modern statistics and wish the reader a "happy bootstrapping".

Notations

This list gives in alphabetical order the symbols that are frequently used throughout this book. Special notation that is used less frequently will be defined as needed.

General notation

A scalar

$\boldsymbol{A}$ column vector or matrix

$\boldsymbol{A}'$ transpose of a vector or a matrix $\boldsymbol{A}$

$\overline{\boldsymbol{A}}$ complex conjugate of a vector or a matrix

$\boldsymbol{A}^H$ Hermitian operation (transpose and complex conjugate) on a vector or a matrix

$\boldsymbol{A}^{-1}$ inverse of a matrix

$\|\boldsymbol{A}\|$ Euclidean vector norm

$|A|$ magnitude of A

$\lfloor A \rfloor$ largest integer $\leq A$

$\lceil A \rceil$ largest integer $\geq A$

$\widehat{\boldsymbol{A}}$ estimator or estimate of $\boldsymbol{A}$

j imaginary unit, $j^2 = -1$

E expectation operator

mod modulo operator

Prob probability

Prob_* probability conditioned on observed data

$\tanh$ hyperbolic tangent

var variance operation

Latin symbols

$c_{XX}(t)$ covariance function of a stationary signal X_t

$f_{XX}(\omega)$ spectral density of a stationary signal X_t

F distribution function

h kernel width or bandwidth

$\boldsymbol{I}$ identity matrix

$I_{XX}(\omega)$ periodogram of an observed stationary signal X_t

k discrete frequency parameter

$K(\cdot)$ kernel function

n size of a random sample

$N(\mu, \sigma^2)$ Gaussian distribution with mean μ and variance σ^2

$o(\cdot)$ order notation: "of smaller order than"

$O(\cdot)$ order notation: "of the same order as"

P_D probability of detection

P_F probability of false alarm

$\mathbb{R}$ the set of real numbers

t discrete or continuous time index

t_n t-distribution with n degrees of freedom

T_n test statistic

$U(a, b)$ uniform distribution over $[a, b]$.

X_t random signal

$\mathcal{X}$ random sample

$\mathcal{X}^*$ bootstrap resample of $\mathcal{X}$

$\mathbb{Z}$ the set of integers

Greek symbols

α level of significance

$\delta(\cdot)$ Kronecker's delta function

χ_n^2 chi-square distribution with n degrees of freedom

Γ mean-square prediction error

θ parameter

μ mean

τ time delay or lag

σ^2 variance

$\Phi(x)$ standard Gaussian distribution function

ω radial frequency

Acronyms

AIC	Akaike information criterion
AR	autoregression
CDF	cumulative distribution function
CFAR	constant false alarm rate
CFD	central finite difference
FIR	finite impulse response
GPR	ground penetrating radar
GPS	global positioning system
HF	high frequency
IF	instantaneous frequency
iid	independent and identically distributed
MDL	minimum distance length (criterion)
MISO	multiple input single output
MLE	maximum likelihood estimator/estimation
ROC	receiver operating characteristic
SNR	signal-to-noise ratio
SRB	sequentially rejective Bonferroni (procedure)
UMP	uniformly most powerful (test)

1

Introduction

Signal processing has become a core discipline in engineering research and education. Many modern engineering problems rely on signal processing tools. This could be either for filtering the acquired measurements in order to extract and interpret information or for making a decision as to the presence or absence of a signal of interest. Generally speaking, statistical signal processing is the area of signal processing where mathematical statistics is used to solve signal processing problems. Nowadays, however, it is difficult to find an application of signal processing where tools from statistics are not used. A statistician would call the area of statistical signal processing time series analysis.

In most statistical signal processing applications where a certain parameter is of interest there is a need to provide a rigorous statistical performance analysis for parameter estimators. An example of this could be finding the accuracy of an estimator of the range of a flying aircraft in radar. These estimators are usually computed based on a finite number of measurements, also called a sample. Consider, for example, a typical radar scenario, in which we aim to ascertain whether the received signal contains information about a possible target or is merely interference. The decision in this case, based on calculating the so-called test statistic, has to be supported with statistical measures, namely the probability of detection and the probability of false alarm. Such a decision can be made if the distribution of the test statistic is known in both cases: when the received signal contains target information and when the target information is absent.

Another important problem in signal processing is to make certain probability statements with respect to a true but unknown parameter. For example, given some estimator of a parameter, we would like to determine upper and lower limits such that the true parameter lies within these limits with a preassigned probability. These limits constitute the so-called confidence interval (Cramér, 1999).

Two main questions arise in a parameter estimation problem. Given a number of measurements and a parameter of interest:

(i) What estimator should we use?

(ii) Having decided to use a particular estimator, how accurate is it?

A signal processing practitioner would first attempt to use the method of maximum likelihood or the method of least squares to answer the first question (Scharf, 1991; Kay, 1993). Having computed the parameter estimate, its accuracy could be measured by the variance of the estimator or a confidence interval for the parameter of interest. In most cases, however, techniques available for computing statistical characteristics of parameter estimators assume that the size of the available set of samples is sufficiently large, so that asymptotic results can be applied. Techniques that invoke the Central Limit Theorem and the assumption of Gaussianity of the noise process are examples of such an approach (Bhattacharya and Rao, 1976; Serfling, 1980).

Let us consider an example where we are interested in finding the 100α, $0 < \alpha < 1$, percent confidence interval for the power spectral density of a stationary real-valued signal, given a finite number of observations. If we were to assume that the number of observations is large, we would use an asymptotic approach so that the spectral density estimates at distinct frequencies could be considered independent with a limiting scaled χ^2 distribution.

Let us mathematically formalise this problem. Assume $X_1, \ldots, X_n$ to be a finite set of observations from a real-valued, strictly stationary signal X_t, $t \in \mathbb{Z}$, with mean zero and a finite variance. Define the spectral density of X_t by

$$C_{XX}(\omega) = \frac{1}{2\pi} \sum_{\tau=-\infty}^{\infty} \mathsf{E}\left[X_t X_{t-|\tau|}\right] e^{-j\omega\tau}, \tag{1.1}$$

where $\mathsf{E}[\cdot]$ denotes expectation, and let

$$I_{XX}(\omega) = \frac{1}{2\pi n} \left| \sum_{t=1}^{n} X_t e^{-j\omega t} \right|^2, \qquad -\pi < \omega \le \pi, \tag{1.2}$$

denote the periodogram of the sample (Brillinger, 1981; Marple Jr, 1987).

Consider estimating the spectral density $C_{XX}(\omega)$ by a kernel spectral density estimate (the smoothed periodogram), given by

$$\hat{C}_{XX}(\omega; h) = \frac{1}{n\,h} \sum_{k=-M}^{M} K\left(\frac{\omega - \omega_k}{h}\right) I_{XX}(\omega_k), \qquad -\pi < \omega \leq \pi, \quad (1.3)$$

where the kernel $K(\cdot)$ is a symmetric, nonnegative function on the real line, h is its bandwidth, and M is the largest integer less than or equal to $n/2$. Let the discrete frequencies ω_k be given by

$$\omega_k = 2\pi k/n, \qquad -M \leq k \leq M.$$

A variety of kernels can be used in Equation (1.3) but let us choose the Bartlett-Priestley window (Priestley, 1981, p. 444) for $K(\cdot)$. Given the estimate of the power spectral density (1.3), one can approximate its distribution asymptotically, as $n \to \infty$ by

$$C_{XX}(\omega)\,\chi^2_{4m+2}/(4\,m+2),$$

where $m = \lfloor(h\,n - 1)/2\rfloor$, $\lfloor \cdot \rfloor$ denotes the floor operator and χ^2_{4m+2} is the χ^2 distribution with $4\,m+2$ degrees of freedom. This leads to the following 100α percent confidence interval (Brillinger, 1981)

$$\frac{(4\,m+2)\,\hat{C}_{XX}(\omega; h)}{\chi^2_{4m+2}\left(\dfrac{1+\alpha}{2}\right)} < C_{XX}(\omega) < \frac{(4\,m+2)\,\hat{C}_{XX}(\omega; h)}{\chi^2_{4m+2}\left(\dfrac{1-\alpha}{2}\right)}, \qquad (1.4)$$

where $\chi^2_\nu(\alpha)$ denotes a number such that the probability

$$\mathsf{Prob}\left[\chi^2_\nu < \chi^2_\nu(\alpha)\right] = \alpha.$$

The analytical result in (1.4) is pleasing, but it assumes that n is sufficiently large so that $\hat{C}_{XX}(\omega_1),\dots,\hat{C}_{XX}(\omega_M)$ are independent χ^2 random variables. In many signal processing problems, as will be seen throughout the book, large sample methods are inapplicable. This is either because of time constraints or because the signal of interest is non-stationary and stationarity can be assumed over a small portion of data only.

There are cases where small sample results, obtained analytically, do exist (Shenton and Bowman, 1977; Field and Ronchetti, 1990). However, more often it is the case that these results cannot be attained and one may have to resort to Monte Carlo simulations (Robert and Casella, 1999). We will recall the problem of estimating the 100α percent confidence interval for power spectral densities in Chapter 2, where a bootstrap-based solution is described.

The bootstrap was introduced by Bradley Efron (1979a,b, 1981, 1982) more than two decades ago, mainly to calculate confidence intervals for parameters in situations where standard methods were not applicable (Efron and Gong, 1983). An example of this would be a situation where the number of observations is so small that asymptotic results are unacceptably inaccurate. Since its invention, the bootstrap has seen many more applications and has been used to solve problems which would be too complicated to be solved analytically (Hall, 1992; Efron and Tibshirani, 1993). Before we continue, let us clarify several questions that we have been frequently asked in the past.

What is the bootstrap? Simply put, the bootstrap is a method which does with a computer what the experimenter would do in practice if it were possible: they would repeat the experiment. With the bootstrap, a new set of experiments is not needed, instead, the original data is reused. Specifically, the original observations are randomly reassigned and the estimate is recomputed. These assignments and recomputations are done a large number of times and considered as repeated experiments. One may think that the bootstrap is similar to Monte Carlo simulations. However, this is not the case. The main advantage of the bootstrap over Monte Carlo simulations is that the bootstrap does not require the experiment to be repeated.

From a data manipulation point of view, the main idea encapsulated by the bootstrap is to simulate as much of the "real world" probability mechanism as possible, substituting any unknowns with estimates from the observed data. Through the simulation in the "bootstrap world", unknown entities of interest in the "real world" can be estimated as indicated in Figure 1.1. The technical aspects of such simulation are covered in the next chapter.

Why is the bootstrap so attractive? The bootstrap does not make the assumption of a large number of observations of the signal. It can answer many questions with very little in the way of modelling, assumptions or analysis and can be applied easily. In an era of exponentially declining computational costs, computer-intensive methods such as the bootstrap are becoming a bargain. The conceptual simplicity of bootstrap methods can sometimes undermine the rich and difficult theory upon which they are based (Hall, 1992; Shao and Tu, 1995). In the next chapter, we will provide a review of the bootstrap theory in a manner more accessible to signal processing practitioners.

What can I use the bootstrap for? In general, the bootstrap is a methodology for answering the question we posed earlier, that is, *how accurate is a parameter estimator?* It is a fundamental question in many signal processing problems and we will see later how one, with the bootstrap, can

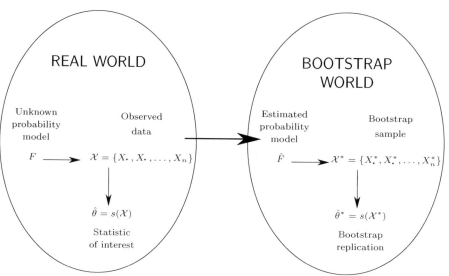

Fig. 1.1. The bootstrap approach, adapted from Efron and Tibshirani (1993, Fig. 8.3). See Chapter 2 for a technical interpretation of this figure.

solve many more problems encountered by a signal processing engineer today, for example, signal detection. This text will also provide an answer to the question regarding the choice of an estimator from among a family of estimators using the bootstrap. We briefly discuss this topic in Chapter 5, where we consider the optimisation of trimming for the trimmed mean in a radar application (see also the work of Léger *et al.* (1992), for example).

Is the bootstrap always applicable? Theoretical work on the bootstrap and applications have shown that bootstrap methods are potentially superior to large sample techniques. A danger, however, does exist. The signal processing practitioner may well be attracted to apply the bootstrap in an application to avoid the use of methods that invoke strong assumptions, such as asymptotic theory, because these are judged inappropriate. But in this case the bootstrap may also fail (Mammen, 1992; Young, 1994). Special care is therefore required when applying the bootstrap in real-life situations (Politis, 1998). The next chapter provides the fundamental concepts and methods needed by the signal processing practitioner to decide when and how to apply the bootstrap successfully.

Applications of bootstrap methods to real-life engineering problems have been reported in many areas, including radar and sonar signal processing, geophysics, biomedical engineering and imaging, pattern recognition and computer vision, image processing, control, atmospheric and environmental research, vibration analysis and artificial neural networks. In almost all these fields, bootstrap methods have been used to approximate the distribution of an estimator or some of its characteristics. Let us list in no particular order some of the bootstrap engineering applications that we found interesting.

Radar and sonar: The bootstrap has been applied to radar and sonar problems for more than a decade. Nagaoka and Amai (1990, 1991) discuss an application in which the distribution of the estimated "close approach probability" is derived to be used as an index of collision risk in air traffic control. Hewer *et al.* (1996) consider a wavelet-based constant false alarm rate (CFAR) detector in which the bootstrap is used to derive the statistics of the detector from lexicographically ordered image vectors. Anderson and Krolik (1998a,b, 1999) use the bootstrap in a hidden Markov model approximation to the ground range likelihood function in an over-the-horizon radar application.

Ong and Zoubir (1999a,b, 2000b, 2003) consider bootstrap applications in CFAR detection for signals in non-Gaussian and correlated interference, while Zoubir *et al.* (1999) apply bootstrap methods to the detection of landmines. Böhme and Maiwald (1994) apply bootstrap procedures to signal detection and location using sensor arrays in passive sonar and to the analysis of seismic data.

Krolik *et al.* (1991) use bootstrap methods for evaluating the performance of source localisation techniques on real sensor array data without precise *a priori* knowledge of true source positions and the underlying data distribution (see also (Krolik, 1994)). Reid *et al.* (1996) employ bootstrap based techniques to determine confidence bounds for aircraft parameters given only a single acoustic realisation, while Bello (1998) uses the bootstrap to calculate cumulative receiver operating characteristic (ROC) curve confidence bounds for sets of side-scan sonar data.

Geophysics: A similar interest in bootstrap methods has taken place in geophysics. Fisher and Hall (1989, 1990, 1991) apply the bootstrap to the problem of deciding whether or not palaeomagnetic specimens sampled from a folded rock surface were magnetised before or after folding occurred. They conclude that the bootstrap method provides the only feasible approach in this common palaeomagnetic problem. Another application of bootstrap methods in palaeomagnetism has been reported by Tauxe *et al.* (1991). Kawano and Higuchi (1995) estimate with the bootstrap the average component in the minimum variance direction in space physics.

Ulrych and Sacchi (1995) propose an extended information criterion based on the bootstrap for the estimation of the number of harmonics actually present in geophysical data. Later, Sacchi (1998) uses the bootstrap for high-resolution velocity analysis.

Lanz *et al.* (1998) perform quantitative error analyses using a bootstrap technique while determining the depth and geometry of a landfill's lower boundary. Mudelsee (2000) uses bootstrap resampling in ramp function regression for quantifying climate transitions, while Rao (2000) uses the bootstrap to assess and improve atmospheric prediction models.

Biomedical engineering: Biomedical signal and image processing has been another area of extensive bootstrap applications. Haynor and Woods (1989) use the bootstrap for estimating the regional variance in emission tomography images. Banga and Ghorbel (1993) introduce a bootstrap sampling scheme to remove the dependence effect of pixels in images of the human retina. Coakley (1996) computes bootstrap expectation in the reconstruction of positron emission tomography images. Locascio *et al.* (1997) use the bootstrap to adjust p-values in multiple significance tests across pixels in magnetic resonance imaging. Maitra (1998) applies the bootstrap in estimating the variance in parametric biomedical images. Verotta (1998) investigates the use of the bootstrap to obtain the desired estimates of variability of system kernel and input estimates, while Bullmore *et al.* (2001) use bootstrap based techniques in the time and wavelet domains in neurophysiological time series analysis. Another interesting application of the bootstrap is reported by Iskander *et al.* (2001) where it is used to find the optimal parametric model for the human cornea. Recently, Chen *et al.* (2002) have employed the bootstrap for aiding the diagnosis of breast cancer in ultrasound images.

Image processing: Bootstrap methods have been widely applied in image processing, pattern recognition and computer vision. Jain *et al.* (1987) apply several bootstrap techniques to estimate the error rate of nearest-neighbour and quadratic classifiers. Hall (1989b) calculates confidence regions for hands in degraded images. Archer and Chan (1996) apply the bootstrap to problems in blind image restoration where they calculate confidence intervals for the true image. Cabrera and Meer (1996) use the bootstrap to eliminate the bias of estimators of ellipses, while Saradhi and Murty (2001) employ the bootstrap technique to achieve higher classification accuracy in handwritten digit recognition.

Control: Bootstrap methods have found applications in statistical control. Dejian and Guanrong (1995) apply bootstrap techniques for estimating the distribution of the Lyapunov exponent of an unknown dynamic system using its time series data. Seppala *et al.* (1995) extend the bootstrap percentile method to include a series of subgroups, which are typically used in assessing process control limits. They show that the method achieves comparatively better control limit estimates than standard parametric methods. Ming and Dong (1997) utilise the bootstrap to construct a prediction interval for future observations from a Birnbaum–Saunders distribution that is used as a failure time model. Jones and Woodall (1998) use the bootstrap in control chart procedures, while Aronsson *et al.* (1999), apply bootstrap techniques to control linear stochastic systems. They derive the optimal future control signal so that the unknown noise distribution and uncertainties in parameter estimates are taken into account. Recently, Tjärnström and Ljung (2002) have used the bootstrap to estimate the variance of an undermodelled structure that is not flexible enough to describe the underlying control system without the need for Monte Carlo simulations.

Environmental engineering: Bootstrap techniques have found several applications in atmospheric and environmental research. Hanna (1989) uses the related jackknife procedure and the bootstrap for estimating confidence limits for air quality models. The resampling procedures have been applied to predictions by a number of air quality models for the Carpentaria coastal dispersion experiment. Downton and Katz (1993) use the bootstrap to compute confidence intervals for the discontinuity in variance of temperature time series, while Krzyscin (1997) infers about the confidence limits of the trend slope and serial correlation coefficient estimates for temperature using the bootstrap.

Artificial neural networks: Bootstrap techniques have also been applied in the area of artificial neural networks. Bhide *et al.* (1995) demonstrate the use of bootstrap methods to estimate a distillation process bottoms' composition. Tibshirani (1996) discusses a number of methods for estimating the standard error of predicted values from a multilayered perceptron. He finds that bootstrap methods perform best, partly because they capture variability due to the choice of starting weights. LeBaron and Weigend (1998) use the bootstrap to compare the uncertainty in the solution stemming from the data splitting with neural-network specific uncertainties in application to financial time series. Franke and Neumann (2000) investigate the bootstrap methods in the context of artificial neural networks used for estimating a mapping from an input to an output space. Dupret and Koda (2001) use the bootstrap in neural network based supervised learning to correctly learn a classification problem while White and Racine (2001) employ bootstrap techniques to permit valid statistical inference based on estimated feedforward neural-network models. Recently, Papadopoulos *et al.* (2001) have performed a comparative analysis of confidence estimation methods for neural networks, including the bootstrap.

Other signal processing related areas: Zoubir and Böhme (1995) apply bootstrap techniques to construct multiple hypotheses tests for finding optimal sensor locations for knock detection in spark ignition engines. Bootstrap techniques have also been applied to non-stationary data analysis. Zoubir *et al.* (1994a) use the bootstrap to determine confidence bounds for the instantaneous frequency. The concept was followed by Kijewski and Kareem (2002) to the estimation of structural damping. Abutaleb (2002) considers the bootstrap in phase unwrapping and the estimation of time-varying frequency. Other signal processing applications of the bootstrap can be found in the proceedings of ICASSP-94 Special Session (1994), see also Hero (1996).

This list is by no means complete. Areas such as medicine, pharmacy, chemistry, metrology, nuclear technology, econometrics and financial engineering routinely use bootstrap methods in their statistical analyses.

Also, the list does not include applications of the jackknife such as the work by Thomson and Chave (1991), where the authors approximate confidence intervals for spectra, coherences, and transfer functions for diverse geophysical data. Also, we have not included applications of recent bootstrap derivatives such as the weighted bootstrap (Barbe and Bertail, 1995; Burke, 2000), subsampling techniques (Politis *et al.*, 1999), the smoothed bootstrap (De Angelis and Young, 1992; Guera *et al.*, 1997; Hall and Maesono, 2000), or importance resampling (Hall, 1989a; Do and Hall, 1991; Booth *et al.*, 1993). We will not cover these variants of the bootstrap in the book because we feel that each of these areas deserves a monograph of its own. We are confident that these new bootstrap techniques will play a critical role in the future and will find many applications in signal processing because they do alleviate some of the restrictions imposed by the conventional bootstrap.

Bootstrap techniques have become an off-the-shelf tool in mathematical statistics and standard routines are incorporated in the majority of statistical software packages available in the market place (e.g. SPSS, S-Plus, and StatsDirect). Our objective is to demonstrate the power of bootstrap techniques to signal processing practitioners with the hope that they will solve problems which may be intractable with classical tools.

An outline of the book follows. Chapter 2 has been written with great care. It is the introductory chapter to the bootstrap and is intended for readers new to the bootstrap world. We have provided numerous examples and discussions. Only in this chapter have we embedded MATLAB code so as to make sure the concept of resampling both independent and dependent data is well understood.

Chapter 3 deals with bootstrap statistical testing. This is an important area for a signal processing practitioner dealing with the detection of signals. We have introduced alternatives to classical signal detection when the interference is non-Gaussian or unknown. The bootstrap matched filter or the CFAR bootstrap matched filter are powerful tools for detecting known signals in interference of unknown distribution. We extended the concept to tolerance interval bootstrap matched filters where we ensure that the probability of false alarm is maintained with a preset probability.

Chapter 4 concerns model selection with the bootstrap. Here, we report methods used to select a linear model as well as models of a more complicated structure such as autoregressive processes. We give comparisons with Akaike's information criterion and Rissanen's minimum description length. We report on the detection of the number of sources impinging on a uniform linear array. With few assumptions the methods reported show superior performance compared to classical techniques, including the sphericity test.

Chapter 5 is a collection of interesting real world problems the authors have worked on over the years. We report results in the areas of vibration analysis, sonar, radar, and biomedical engineering. In all applications there was no better way to approach the problem, than with the bootstrap. This is followed by two Appendices. Appendix 1 includes MATLAB code for many of the examples in the book while Appendix 2 contains the Bootstrap MATLAB Toolbox, a set of functions purposely written for this book.

The reader new to the bootstrap world is encouraged to carefully read Chapter 2. Chapters 3 and 4 are two main applications of the bootstrap to classical problems which invoke strong assumptions. The reader interested in signal detection/hypothesis testing is referred to Chapter 3 while one interested in model selection should focus on Chapter 4, they may be read independently of each other. Both chapters also contain a theoretical treatment plus examples and applications. The engineer interested in discovering the power of the bootstrap in real-life applications can skip the theoretical treatments in Chapters 3 and 4 and go directly to Chapter 5.

2

The bootstrap principle

In this chapter we introduce the principle of the bootstrap, provide a re-view of basic resampling techniques, and show how the bootstrap can be used to evaluate the distribution of a parameter estimator. We start with non-parametric and parametric bootstrap resampling techniques, which are essentially designed for independent and identically distributed data, fol-lowed by bootstrap methods for dependent data. Then, we discuss boot-strap pivotal statistics, the nested (double) bootstrap, and the method of variance stabilisation. We comment on the limitations of the bootstrap, provide some guidance for the application of bootstrap methods in practical situations, and show an example of bootstrap failure. Finally, we sketch other trends in bootstrap resampling methodology.

2.1 The principle of resampling

Let $\mathcal{X} = \{X_1, X_2, \ldots, X_n\}$ be a sample, i.e., a collection of n numbers drawn at random from a completely unspecified distribution F. When we say "at random" we mean that the X_i's are *independent and identically distributed* (iid) random variables, each having distribution F. Let θ denote an unknown characteristic of F. It could be the mean or variance of F or even the spectral density function discussed earlier. The problem we wish to solve is to find the distribution of $\hat{\theta}$, an estimator of θ, derived from the sample $\mathcal{X}$. This is of great practical importance as we need to infer θ based on $\hat{\theta}$. For example, if θ is $C_{XX}(\omega)$ defined in (1.1) we would want to test whether $C_{XX}(\omega)$ at a given ω is zero or whether it exceeds a certain bound, based on the estimate constructed from the observations x_t, $t = 1, \ldots, n$ of the stationary process X_t, $t \in \mathbb{Z}$.

One way to obtain the distribution of $\hat{\theta}$ is to repeat the experiment a sufficient number of times and approximate the distribution of $\hat{\theta}$ by the so obtained empirical distribution. This is equivalent to Monte Carlo simulations (Robert and Casella, 1999). In many practical situations, however, this is inapplicable for cost reasons or because the experimental conditions are not reproducible. In some applications, samples may be unavailable, for example, when monitoring vibration signals in a cracking concrete beam.

The bootstrap suggests that we resample from a distribution chosen to be close to F in some sense. This could be the sample (or empirical) distribution $\hat{F}$, which is that probability measure that assigns to a set A in the sample space of X a measure equal to the proportion of sample values that lie in A. It is known that under some mild assumptions $\hat{F}$ approaches F as $n \longrightarrow \infty$ (Hall, 1992). Resampling from $\hat{F}$ is referred to as the *non-parametric bootstrap*.† The idea of resampling from $\hat{F}$ was introduced earlier as a transition from a "real world" to "bootstrap world" (see Figure 1.1). The principle of the non-parametric bootstrap is highlighted in Table 2.1.

Table 2.1. *Principle of the non-parametric bootstrap.*

Step 0. Conduct the experiment to obtain the random sample
$$\mathcal{X} = \{X_1, X_2, \ldots, X_n\}$$
and calculate the estimate $\hat{\theta}$ from the sample $\mathcal{X}$.

Step 1. Construct the empirical distribution $\hat{F}$, which puts equal mass $1/n$ at each observation
$$X_1 = x_1, X_2 = x_2, \ldots, X_n = x_n$$

Step 2. From $\hat{F}$, draw a sample
$$\mathcal{X}^* = \{X_1^*, X_2^*, \ldots, X_n^*\},$$
called the bootstrap resample.

Step 3. Approximate the distribution of $\hat{\theta}$ by the distribution of $\hat{\theta}^*$ derived from the bootstrap resample $\mathcal{X}^*$.

Although Table 2.1 states that we construct $\hat{F}$ and we resample from $\hat{F}$, in practice we do not have to directly estimate F. This means that the bootstrap resampling does not involve explicit calculation of the empirical distribution function, the histogram, or the empirical characteristic function. Also, there is no need to estimate the mean, variance or the higher order moments of F. With the non-parametric bootstrap, we simply use the random sample $\mathcal{X} = \{X_1, X_2, \ldots, X_n\}$ and generate a new sample by sampling with replacement from $\mathcal{X}$. We call this new sample a bootstrap sample

† The *parametric* bootstrap is introduced in Section 2.1.3.

or a bootstrap resample. A practical implementation of the non-parametric bootstrap is given in Figure 2.1.

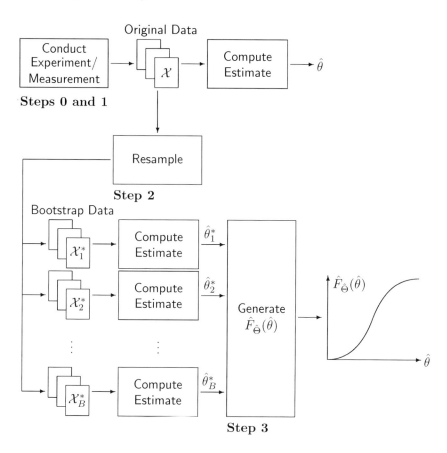

Fig. 2.1. Principle of the non-parametric bootstrap for estimating the distribution function $\hat{F}_{\hat{\Theta}}(\hat{\theta})$ of a parameter estimator $\hat{\theta}(\mathbf{x})$.

Herein, we create a number B of resamples $\mathcal{X}_1^*, \ldots, \mathcal{X}_B^*$. A resample $\mathcal{X}^* = \{X_1^*, X_2^*, \ldots, X_n^*\}$ is an unordered collection of n sample points drawn randomly from $\mathcal{X}$ with replacement, so that each X_i^* has probability n^{-1} of being equal to any one of the X_j's. In other terms,

$$\mathsf{Prob}\left[X_i^* = X_j \,|\, \mathcal{X}\right] = n^{-1}, \qquad 1 \le i, j \le n.$$

That is, the X_i^*'s are independent and identically distributed, conditional on the random sample $\mathcal{X}$, with this distribution (Hall, 1992). This means that $\mathcal{X}^*$ is likely to contain repeats. As an example, consider $n = 4$ and the sample $\mathcal{X} = \{-3.7, 0.5, 1.4, 2.8\}$. The collection $\mathcal{X}^* = \{0.5, -3.7, -3.7, 2.8\}$ should not be mistaken for the set $\{0.5, -3.7, 2.8\}$ because the bootstrap sample has one repeat. Also, $\mathcal{X}^*$ is the same as $\{0.5, -3.7, 2.8, -3.7\}$, $\{-3.7, 0.5, 2.8, -3.7\}$, etc. because the order of elements in the resample plays no role (Hall, 1992; Efron and Tibshirani, 1993). Example 2.1.1 highlights the resampling procedure in a non-parametric bootstrap.

A typical question that arises with the bootstrap is the following: how many resamples does one need? This number depends on the particular application and what is to be estimated. We will discuss this question later.

The resampling procedure is fundamental to the non-parametric bootstrap and needs to be properly understood. To facilitate this, we have written a MATLAB routine called bootrsp.m. Most of the MATLAB functions that were written for this book are collected in Appendix 2 which contains the Bootstrap MATLAB Toolbox. Unless otherwise stated, all routines presented in this book can run with a standard MATLAB application and without the need of additional MATLAB toolboxes. Consider the following simple example of a non-parametric bootstrap resampling procedure.

Example 2.1.1 Non-parametric resampling.

```
>> X=randn(5,1);
>> X'

   X =

      -0.43 -1.66  0.12  0.28 -1.14

>> X_star=bootrsp(X,10)

   X_star =

   -1.14   0.28   0.28   0.12  -0.43  -1.66  -0.43   0.12  -1.14   0.12
   -1.66   0.12   0.28  -1.14  -1.66  -0.43   0.28  -1.14  -0.43   0.28
    0.28  -0.43  -1.14  -1.14  -1.14   0.28   0.12   0.12   0.28   0.12
    0.12  -1.14   0.28   0.12  -0.43  -1.66  -1.14  -1.66  -1.66  -1.66
   -1.14   0.12  -0.43  -1.14  -0.43  -0.43   0.12   0.28  -1.14  -0.43
```

Similarly, we can use bootrsp.m to create resamples of two-dimensional data.

```
>> X=randn(2,3)
```

```
X =

  1.19 -0.03  0.17
  1.18  0.32 -0.18

>> X_star=bootrsp(X,3)

  X_star(:,:,1) =

  1.18  1.18  1.19
  0.17  0.32  0.17

  X_star(:,:,2) =

  -0.03 -0.18 -0.03
  -0.18  0.32 -0.18

  X_star(:,:,3) =

  0.17  0.17 -0.03
  0.32  0.32  1.18
```

With the routine bootrsp.m we can generate an arbitrary number of boot-strap resamples from the given sample $\mathcal{X} = \{X_1, X_2, \ldots, X_n\}$. Some of the values from the original sample will appear in the resample more than once while other values may not appear at all. The fourth resample of X (fourth column of X_star in the first part of this example, has the number -1.14 appearing three times while the numbers $-0.43, -1.66$ and 0.28 do not appear at all. We emphasise that, in general, the size of the bootstrap resample has to be equal to that of the original sample. These are the basics of non-parametric bootstrap resampling. Note, however, that in some boot-strap applications such as the bootstrap model selection (see Chapter 4), subsampling techniques can be used.

At times, we will refer to resampling the data using the bootstrap simply as *bootstrapping*. Bootstrap methods are not limited to resampling uni-variate data and can cater for multivariate data as it was shown in Exam-ple 2.1.1. An example of bootstrapping bi-variate data is given below.

Example 2.1.2 Non-parametric bi-variate resampling.

```
>> X1=randn(4,1);
>> X1'

   X1 =

   0.72 -0.58  2.18 -0.13

>> X2=randn(4,1);
>> X2'

   X2 =

   0.11  1.06  0.05 -0.09

>> [X1_star,X2_star]=bootrsp2(X1,X2,10)

   X1_star =

   -0.58  2.18 -0.13 -0.13 -0.58 -0.13 -0.58 -0.58  2.18  2.18
    0.72  2.18  2.18  0.72 -0.58 -0.13 -0.58 -0.13 -0.58  0.72
    2.18  0.72 -0.58  2.18 -0.13 -0.13  0.72  2.18 -0.58 -0.13
   -0.13 -0.58  2.18  0.72  0.72 -0.13  2.18 -0.58 -0.13  2.18

   X2_star =

    1.06  0.05 -0.09 -0.09  1.06 -0.09  1.06  1.06  0.05  0.05
    0.11  0.05  0.05  0.11  1.06 -0.09  1.06 -0.09  1.06  0.11
    0.05  0.11  1.06  0.05 -0.09 -0.09  0.11  0.05  1.06 -0.09
   -0.09  1.06  0.05  0.11  0.11 -0.09  0.05  1.06 -0.09  0.05
```

The bi-variate resampling has been performed using the bootrsp2.m routine (see Appendix 2). A simple analysis of this routine indicates that multivariate resampling is performed using the same bootstrap index for all variables. Take, for example, the first columns of X1_star and X2_star in Example 2.1.2. It is clear that the resample consists of the second and first elements followed by the third and fourth elements of the original samples X1 and X2.

A reliable pseudo-random number generator is essential for valid application of the bootstrap method. In all our applications, we use a pseudo-random number generator proposed by Park and Miller (1988), which is a built-in routine in MATLAB. More detailed treatment of pseudo-random number generators can be found in the works of Knuth (1981) and Bratley *et al.* (1983).

2.1.1 Some theoretical results for the mean

Examples 2.1.1 and 2.1.2 give a general idea about bootstrap resampling. The question we may pose now is how much information from the original sample is retained in the bootstrap resample. To answer this question we need to evaluate the consistency and convergence of bootstrap resampling. In the statistical literature, consistency and convergence results have been reported for many types of statistics. Let us consider the following three results for a simple statistic, the sample mean.

Result 2.1.1 Bootstrap convergence for the sample mean.
Given a random sample $\mathcal{X} = \{X_1, \ldots, X_n\}$, consider the parameter of interest to be $\theta = \mathsf{E}[X]$. An obvious estimator for θ is the sample mean

$$\hat{\theta} = \frac{1}{n} \sum_i X_i.$$

Let its bootstrap equivalent be

$$\hat{\theta}^* = \frac{1}{n} \sum_i X_i^*.$$

If $\mathsf{E}[X^2] < \infty$, then (Singh, 1981, Eq. (1.1))

$$\sup_x \left| \mathsf{Prob}_* [\sqrt{n}(\hat{\theta}^* - \hat{\theta}) \leq x] - \mathsf{Prob}[\sqrt{n}(\hat{\theta} - \theta) \leq x] \right| \to 0 \quad a.s.$$

where $\mathsf{Prob}_*[\cdot]$ is the bootstrap probability conditioned on the observed data. Setting $T_n^* = \sqrt{n}(\hat{\theta}^* - \hat{\theta})$ and $T_n = \sqrt{n}(\hat{\theta} - \theta)$, the result above proves that the distribution of T_n^* converges to that of T_n in the sense that the sup-norm of the difference is almost surely (a.s.) zero.

Similar results can be derived for a standardised and studentised sample mean. These two results are of importance to bootstrap hypothesis testing which we will discuss in Chapter 3.

Result 2.1.2 Bootstrap convergence for the standardised sample mean.

Let θ, $\hat{\theta}$ and $\hat{\theta}^*$ represent the mean, the sample mean, and the bootstrap sample mean as in Result 2.1.1. Let $\sigma^2 = \text{var}[X]$, and

$$\hat{\sigma}^2 = \frac{1}{n}\sum_i (X_i - \hat{\theta})^2$$

be the biased sample variance. If $\mathsf{E}[|X_1|^3] < \infty$ and F is not a lattice distribution,† then (Singh, 1981, Eq. (1.5))

$$\sqrt{n}\sup_x \left| \mathsf{Prob}_*[\sqrt{n}(\hat{\theta}^* - \hat{\theta})/\hat{\sigma} \le x] - \mathsf{Prob}[\sqrt{n}(\hat{\theta} - \theta)/\sigma \le x] \right| \to 0 \quad a.s.$$

If the distribution of $T_n = \sqrt{n}(\hat{\theta} - \theta)/\sigma$ is approximated by the standard Gaussian, the sup-norm distance is $O(n^{-1/2})$ compared with $o(n^{-1/2})$ if it is approximated by the distribution of $T_n^* = \sqrt{n}(\hat{\theta}^* - \hat{\theta})/\hat{\sigma}$ (Shao and Tu, 1995, p. 93). This highlights the higher-order accuracy of the bootstrap distribution.

Result 2.1.3 Bootstrap convergence for the studentised sample mean.

Let θ, $\hat{\theta}$ and $\hat{\theta}^*$ represent the mean, the sample mean, and the bootstrap sample mean as in Result 2.1.1. Let $\hat{\sigma}^2$ be the biased sample variance as defined in Result 2.1.2, and

$$\hat{\sigma}^{*2} = \frac{1}{n}\sum_i (X_i^* - \hat{\theta}^*)^2.$$

If $\mathsf{E}[|X_1|^6] < \infty$, then

$$\sqrt{n}\sup_x \left| \mathsf{Prob}_*[\sqrt{n}(\hat{\theta}^* - \hat{\theta})/\hat{\sigma}^* \le x] - \mathsf{Prob}[\sqrt{n}(\hat{\theta} - \theta)/\hat{\sigma} \le x] \right| \to 0 \quad a.s.$$

This result is a special case of the result by Babu and Singh (1983) for studentised linear combinations of sample means. It states that the sup-norm difference between the distributions of the studentised sample mean statistic $T_n = \sqrt{n}(\hat{\theta} - \theta)/\hat{\sigma}$ and the bootstrap statistic $T_n^* = \sqrt{n}(\hat{\theta}^* - \hat{\theta})/\hat{\sigma}^*$ is of order $o(n^{-1/2})$ (Ong, 2000).

Convergence and consistency results can be found for many statistics other than the mean and the interested reader is referred to the statistical literature on this topic, see for example (Shao and Tu, 1995). The important message here is that, generally speaking, the bootstrap resamples contain

---

† F does not have all atoms of the form $x_\bullet + k\Delta x$ for integer k and some constants $x_\bullet$ and Δx.

the same statistical information as the original sample, provided that the data is iid.

2.1.2 Examples of non-parametric bootstrap estimation

A common use of the non-parametric bootstrap is to estimate the distribution of a parameter estimator from a random sample. Thus, quantities such as bias, variance, or confidence bands can also be easily estimated. To get a better understanding of a non-parametric bootstrap, we will consider the following examples.

Example 2.1.3 Bias estimation.

Consider the problem of estimating the variance of an unknown distribution with parameters μ and σ which we denote by $F_{\mu,\sigma}$, given the random sample $\mathcal{X} = \{X_1, \ldots, X_n\}$. Two different estimators are commonly used:

$$\hat{\sigma}_u^2 = \frac{1}{n-1} \sum_{i=1}^{n} \left(X_i - \frac{1}{n} \sum_{j=1}^{n} X_j \right)^2 \tag{2.1}$$

and

$$\hat{\sigma}_b^2 = \frac{1}{n} \sum_{i=1}^{n} \left(X_i - \frac{1}{n} \sum_{j=1}^{n} X_j \right)^2. \tag{2.2}$$

It can easily be shown that

$$\mathsf{E}[\hat{\sigma}_u^2] = \sigma^2 \quad \text{and} \quad \mathsf{E}[\hat{\sigma}_b^2] = \left(1 - \frac{1}{n} \right) \sigma^2.$$

Thus, $\hat{\sigma}_b^2$ is a biased estimator for σ^2 while $\hat{\sigma}_u^2$ is an unbiased estimator of σ^2. With the bootstrap, one can estimate the bias

$$b(\hat{\sigma}^2) = \mathsf{E}[\hat{\sigma}^2 - \sigma^2]$$

by

$$\mathsf{E}_*[\hat{\sigma}^{*2} - \hat{\sigma}^2],$$

where $\hat{\sigma}^2$ is the maximum likelihood estimate of σ^2 and $\mathsf{E}_*[\cdot]$ is expectation with respect to bootstrap sampling. Note that in the Gaussian case the maximum likelihood estimate of σ^2 is $\hat{\sigma}_b^2$. It should also be noted that the bootstrap variance $\hat{\sigma}^{*2}$ is strongly consistent (Shao and Tu, 1995, p. 87). We can adapt the steps of the non-parametric bootstrap from Table 2.1 and write a bootstrap procedure for bias estimation as illustrated in Table 2.2.

Table 2.2. *Bootstrap algorithm for bias estimation.*

Step 0. *Experiment.* Collect the data into

$$\mathcal{X} = \{X_1, \ldots, X_n\}$$

and compute the estimates $\hat{\sigma}_u^2$ and $\hat{\sigma}_b^2$ according to (2.1) and (2.2).

Step 1. *Resampling.* Draw a random sample of size n, with replacement, from $\mathcal{X}$.

Step 2. *Calculation of the bootstrap estimate.* Calculate the bootstrap estimates $\hat{\sigma}_u^{*2}$ and $\hat{\sigma}_b^{*2}$ from $\mathcal{X}^*$ in the same way $\hat{\sigma}_u^2$ and $\hat{\sigma}_b^2$ were computed but with the resample $\mathcal{X}^*$.

Step 3. *Repetition.* Repeat Steps 1 and 2 to obtain a total of B bootstrap estimates $\hat{\sigma}_{u,1}^{*2}, \ldots, \hat{\sigma}_{u,B}^{*2}$ and $\hat{\sigma}_{b,1}^{*2}, \ldots, \hat{\sigma}_{b,B}^{*2}$.

Step 4. *Bias estimation.* Estimate $b(\hat{\sigma}_u^2)$ by

$$b_*(\hat{\sigma}_u^{*2}) = \frac{1}{B} \sum_{i=1}^{B} \hat{\sigma}_{u,i}^{*2} - \hat{\sigma}_b^2$$

and $b(\hat{\sigma}_b^2)$ by

$$b_*(\hat{\sigma}_b^{*2}) = \frac{1}{B} \sum_{i=1}^{B} \hat{\sigma}_{b,i}^{*2} - \hat{\sigma}_b^2.$$

Let us consider a sample of size $n = 5$ from the standard Gaussian distribution. With $B = 1000$ and 1000 Monte Carlo simulations, we obtained the histograms for $b_*(\hat{\sigma}_u^{*2})$ and $b_*(\hat{\sigma}_b^{*2})$ as shown in Figure 2.2. It is worth noting that the sample mean for $b(\hat{\sigma}_u^2)$ was -4.94×10^{-4} as compared to the theoretical value of zero while the sample mean for $b(\hat{\sigma}_b^2)$ was -0.1861 compared to the theoretical value of -0.2. A MATLAB code for the bias estimation example is given in Section A1.1 of Appendix 1.

Note that the sample mean in the above example will converge to the true value of the bias as n increases. Of course the original sample has to be a good representation of the population and it should not include too many outliers.

Estimating the bias of a parameter estimator with the bootstrap can be also used in the context of sharpening an estimator. Kim and Singh (1998) used a bootstrap method to check whether the use of bias correction leads to an increase or a decrease in the mean squared error of the estimator.

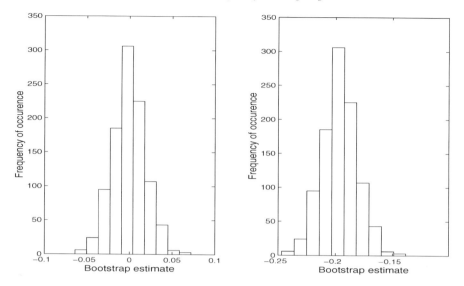

Fig. 2.2. Bootstrap estimation of the bias $b(\hat{\sigma}_u^2)$ (left) and of $b(\hat{\sigma}_b^2)$ (right).

Example 2.1.4 Variance estimation.

Consider now the problem of finding the variance $\sigma_{\hat{\theta}}^2$ of an estimator $\hat{\theta}$ of θ, based on the random sample $\mathcal{X} = \{X_1, \ldots, X_n\}$ from the unknown distribution F_θ.

In some cases one may derive an analytic expression for $\sigma_{\hat{\theta}}^2$ if this is mathematically tractable. Alternatively, one may use asymptotic arguments to compute an estimate $\hat{\sigma}_{\hat{\theta}}^2$ for $\sigma_{\hat{\theta}}^2$. However, there are many situations where the conditions for the above are not fulfilled. In the example of the spectral density $C_{XX}(\omega)$ in Equation (1.1) an estimate of the variance for $\hat{C}_{XX}(\omega)$ is available only for n large. One way to overcome the problem is to use the bootstrap to approximate $\sigma_{\hat{\theta}}^2$ by $\hat{\sigma}_{\hat{\theta}}^{*2}$. Similarly to the previous example, we will develop a bootstrap algorithm in which we will use the original sample $\mathcal{X} = \{X_1, \ldots, X_n\}$ to generate several bootstrap resamples. The resamples will then be used to derive the bootstrap estimates used to calculate the estimate $\hat{\sigma}_{\hat{\theta}}^{*2}$. The procedure is shown in Table 2.3.

Suppose that the distribution $F_{\mu,\sigma}$ is Gaussian with mean $\mu = 10$ and variance $\sigma^2 = 25$, and we wish to estimate $\sigma_{\hat{\mu}}$ based on a random sample $\mathcal{X}$ of size $n = 50$. Following the procedure of Table 2.3 with an appropriate number of bootstrap resamples (Efron and Tibshirani, 1993) for variance estimation, $B = 25$, a bootstrap estimate of the variance of $\hat{\mu}$ is found to be $\hat{\sigma}_{\text{BOOT}}^2 = 0.49$ as compared to the true $\sigma_{\hat{\mu}}^2 = \sigma^2/n = 0.5$. A typical histogram

Table 2.3. *Bootstrap algorithm for variance estimation.*

Step 0. *Experiment.* Conduct the experiment and collect the random data into the sample

$$\mathcal{X} = \{X_1, \ldots, X_n\}.$$

Step 1. *Resampling.* Draw a random sample of size n, with replacement, from $\mathcal{X}$.

Step 2. *Calculation of the bootstrap estimate.* Evaluate the bootstrap estimate $\hat{\theta}^*$ from $\mathcal{X}^*$ calculated in the same manner as $\hat{\theta}$ but with the resample $\mathcal{X}^*$ replacing $\mathcal{X}$.

Step 3. *Repetition.* Repeat Steps 1 and 2 many times to obtain a total of B bootstrap estimates $\hat{\theta}_1^*, \ldots, \hat{\theta}_B^*$.

Step 4. *Estimation of the variance of $\hat{\theta}$.* Estimate the variance $\sigma_{\hat{\theta}}^2$ of $\hat{\theta}$ by

$$\hat{\sigma}_{\text{BOOT}}^2 = \frac{1}{B-1} \sum_{i=1}^{B} \left(\hat{\theta}_i^* - \frac{1}{B} \sum_{j=1}^{B} \hat{\theta}_j^* \right)^2.$$

---

of $\hat{\sigma}_{\hat{\mu}}^{*2(1)}, \hat{\sigma}_{\hat{\mu}}^{*2(2)}, \ldots, \hat{\sigma}_{\hat{\mu}}^{*2(1000)}$ is shown in Figure 2.3. A MATLAB routine for this example is given in Section A1.1 of Appendix 1.

Confidence interval estimation is one of the most frequent applications of the bootstrap. It is not fundamentally different from hypothesis testing, which will be discussed in Chapter 3. Because the emphasis in this book is more towards application of the bootstrap for hypothesis testing or signal detection, we will put more emphasis on this example.

Example 2.1.5 Confidence interval for the mean.

Let $X_1, \ldots, X_n$ be n iid random variables from some unknown distribution $F_{\mu,\sigma}$. We wish to find an estimator and a $100(1-\alpha)\%$ interval for the mean μ. Usually, we use the sample mean as an estimate for μ,

$$\hat{\mu} = \frac{1}{n} \sum_{i=1}^{n} X_i.$$

A confidence interval for μ can be found by determining the distribution of $\hat{\mu}$ (over repeated samples of size n from the underlying distribution), and

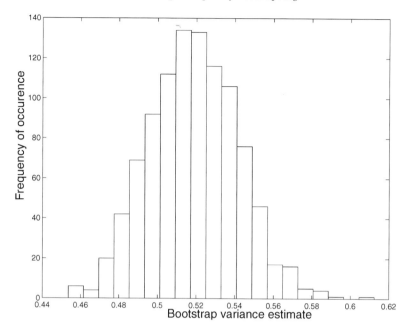

Fig. 2.3. Histogram of $\hat{\sigma}^{*2}_{\hat{\mu},(1)}, \hat{\sigma}^{*2}_{\hat{\mu},(2)}, \ldots, \hat{\sigma}^{*2}_{\hat{\mu},(1000)}$, based on a random sample of size $n = 50$ and $B = 25$.

finding values $\hat{\mu}_L, \hat{\mu}_U$ such that

$$\mathsf{Prob}[\hat{\mu}_L \le \mu \le \hat{\mu}_U] = 1 - \alpha.$$

The distribution of $\hat{\mu}$ depends on the distribution of the X_i's, which is unknown. In the case where n is large the distribution of $\hat{\mu}$ could be approximated by the Gaussian distribution as per the Central Limit Theorem (Kendall and Stuart, 1967; Manoukian, 1986), but such an approximation is not valid in applications where n is small.

As mentioned earlier, the bootstrap assumes that the random sample $\mathcal{X} = \{X_1, \ldots, X_n\}$ itself constitutes the underlying distribution. Then, by resampling from $\mathcal{X}$ many times and computing $\hat{\mu}$ for each of these resamples, we obtain a bootstrap distribution for $\hat{\mu}$ which approximates the distribution of $\hat{\mu}$, and from which a confidence interval for μ is derived. This procedure is described in Table 2.4, where a sample of size 10 is taken from the Gaussian distribution with mean $\mu = 10$, variance $\sigma^2 = 25$, and where the level of confidence is 95%.

We used the same data and algorithm of Table 2.4 with other α values. For example, we found the 99% confidence interval to be $(4.72, 14.07)$ and with $B = 100$ resamples, we found $(7.33, 12.93)$ to be the 90% confidence

Table 2.4. *The bootstrap principle for calculating a confidence interval for the mean.*

Step 0. *Experiment.* Conduct the experiment. Suppose our sample is

$$\mathcal{X} = \{-2.41, 4.86, 6.06, 9.11, 10.20, 12.81, 13.17, 14.10, 15.77, 15.79\}$$

of size 10, with $\hat{\mu} = 9.946$ being the mean of all values in $\mathcal{X}$.

Step 1. *Resampling.* Draw a random sample of 10 values, with replacement, from $\mathcal{X}$. One might obtain the *bootstrap resample*

$$\mathcal{X}^* = \{9.11, 9.11, 6.06, 13.17, 10.20, -2.41, 4.86, 12.81, -2.41, 4.86\}.$$

Note here that some of the original sample values appear more than once, such as 9.11 and others not at all, such as 14.10.

Step 2. *Calculation of the bootstrap estimate.* Calculate the mean of all values in $\mathcal{X}^*$. The obtained mean of all 10 values in $\mathcal{X}^*$ is $\hat{\mu}_1^* = 6.54$.

Step 3. *Repetition.* Repeat Steps 1 and 2 a large number of times to obtain a total of B bootstrap estimates $\hat{\mu}_1^*, \ldots, \hat{\mu}_B^*$. For example, let $B = 1000$.

Step 4. *Approximation of the distribution of $\hat{\mu}$.* Sort the bootstrap estimates into increasing order to obtain $\hat{\mu}_{(1)}^* \leq \hat{\mu}_{(2)}^* \leq \cdots \leq \hat{\mu}_{(1000)}^*$, where $\hat{\mu}_{(k)}^*$ is the kth smallest of $\hat{\mu}_1^*, \ldots, \hat{\mu}_B^*$. For example, we might get

$$3.48, 3.39, 4.46, \ldots, 8.86, 8.89, \ldots, 10.07, 10.08, \ldots, 14.46, 14.53, 14.66.$$

A histogram of the obtained bootstrap estimates $\hat{\mu}_1^*, \ldots, \hat{\mu}_B^*$ is given in Figure 2.4 along with the density function of the Gaussian distribution with mean $\mu = 10$ and variance $\sigma^2/n = 25$.

Step 5. *Confidence interval.* The desired $100(1 - \alpha)\%$ bootstrap confidence interval is $(\hat{\mu}_{(q_1)}^*, \hat{\mu}_{(q_2)}^*)$, where $q_1 = \lfloor B\alpha/2 \rfloor$ is the integer part of $B\alpha/2$ and $q_2 = B - q_1 + 1$. For $\alpha = 0.05$ and $B = 1000$, we get $q_1 = 25$ and $q_2 = 976$, and the 95% confidence interval is found to be $(6.27, 13.19)$ as compared to the theoretical $(6.85, 13.05)$.

---

interval. Note that the number of resamples B has to be chosen according to the desired confidence interval. For example, we could not estimate the 99% interval using $B = 100$ resamples. We will return to this issue in Chapter 3, in the context of hypothesis testing. The MATLAB code for the bootstrap procedure for estimating a confidence interval for the mean and the particular example from Table 2.4 are given in Section A1.1 of Appendix 1.

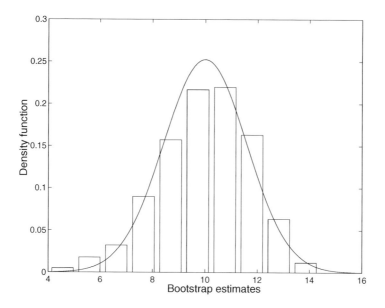

Fig. 2.4. Histogram of $\hat{\mu}_1^*, \hat{\mu}_2^*, \ldots, \hat{\mu}_{1000}^*$ based on the random sample $\mathcal{X} = \{-2.41,$ 4.86, 6.06, 9.11, 10.20, 12.81, 13.17, 14.10, 15.77, 15.79\}. The solid line indicates the probability density function of a Gaussian variable with mean 10 and variance 25.

We also ran the algorithm of Table 2.4 using a random sample of size 10 from the t-distribution with four degrees of freedom. The histogram of the bootstrap estimates so obtained is shown in Figure 2.5. As the theoretical fit in this case is not available, we compare the result with the smoothed empirical density function (or more precisely a kernel density estimate) based on 1000 Monte Carlo replications. In this example we used the Gaussian kernel with the optimum width $h = 1.06 B^{(-1/5)} \hat{\sigma}_0 = 0.12$, where $\hat{\sigma}_0$ is the standard deviation of the estimates of μ, obtained through Monte Carlo replications (Silverman, 1986). The 95 % confidence interval for μ was found to be $(-0.896, 0.902)$ and $(-0.886, 0.887)$ based on the bootstrap and Monte Carlo estimates, respectively.

In practice, the procedure described in Table 2.4 can be substantially improved because the interval calculated is, in fact, an interval with coverage less than the nominal value (Hall, 1988). In Section 2.2, we will discuss another way of constructing confidence intervals for the mean that would lead to a more accurate result. The computational expense to calculate the confidence interval for μ is approximately B times greater than the one

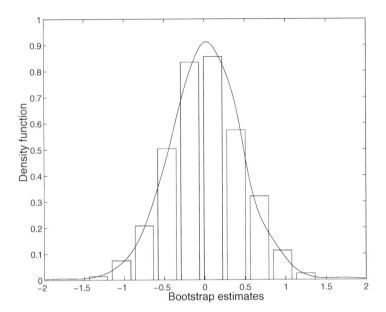

Fig. 2.5. Histogram of 1000 bootstrap estimates of the mean of the t-distribution with four degrees of freedom. The solid line indicates the kernel probability density function obtained from 1000 Monte Carlo simulations.

needed to compute $\hat{\mu}$. However, this is generally acceptable given the ever-increasing capabilities of today's computers.

The three examples provided above give an insight into the basic procedure of the non-parametric bootstrap. We now turn our attention to the parametric bootstrap.

2.1.3 The parametric bootstrap

Suppose that one has partial information of F. For example, F is known to be the Gaussian distribution but with unknown mean μ and variance σ^2. This suggests that we draw a resample of size n from the Gaussian distribution with mean $\hat{\mu}$ and variance $\hat{\sigma}^2$ where $\hat{\mu}$ and $\hat{\sigma}^2$ are estimated from $\mathcal{X}$ rather than a non-parametric estimate $\hat{F}$ of F. With this method, known as the *parametric bootstrap* (Hall, 1992; Efron and Tibshirani, 1993), one hopes to improve upon the non-parametric bootstrap described above.

In general terms, given that the sample $\mathcal{X}$ is from F_θ, we first estimate θ (which could well be a vector of parameters as in the Gaussian case) and draw from $\hat{F}_{\hat{\theta}}$ B bootstrap samples of size n each,

$$\hat{F}_{\hat{\theta}} \longrightarrow (x_1^*, x_2^*, \ldots, x_n^*)$$

The steps of the parametric bootstrap follow immediately. We use a pseudo-random number generator to draw new samples from the distribution type where the unknowns are replaced by their estimates. When used in a parametric way, the bootstrap provides more accurate answers, provided the model is correct.

Example 2.1.6 Confidence interval for the mean: a parametric approach.

To illustrate the concept of a parametric bootstrap, let us consider the problem of finding a confidence interval for the mean as in Example 2.1.5, but this time we will assume that the model is known. The parametric bootstrap procedure for finding a confidence interval is shown in Table 2.5, while the histogram of bootstrap estimates is shown in Figure 2.6.

Table 2.5. *The parametric bootstrap principle for calculating a confidence interval for the mean.*

Step 0. *Experiment.* Conduct the experiment and collect $X_1, \ldots, X_n$ into $\mathcal{X}$. Suppose $F_{\mu,\sigma}$ is $\mathcal{N}(10, 25)$ and

$$\mathcal{X} = \{-2.41, 4.86, 6.06, 9.11, 10.20, 12.81, 13.17, 14.10, 15.77, 15.79\}$$

is of size $n = 10$. The mean of all values in $\mathcal{X}$ is $\hat{\mu} = 9.95$ and the sample variance is $\hat{\sigma}^2 = 33.15$.

Step 1. *Resampling.* Draw a sample of 10 values, with replacement, from $F_{\hat{\mu},\hat{\sigma}}$. We might obtain

$$\mathcal{X}^* = \{7.45, 0.36, 10.67, 11.60, 3.34, 16.80, 16.79, 9.73, 11.83, 10.95\}.$$

Step 2. *Calculation of the bootstrap estimate.* Calculate the mean of $\mathcal{X}^*$. The mean of all 10 values in $\mathcal{X}^*$ is $\hat{\mu}_1^* = 9.95$.

Step 3. *Repetition.* Repeat Steps 1 and 2 to obtain B bootstrap estimates $\hat{\mu}_1^*, \ldots, \hat{\mu}_B^*$. Let $B = 1000$.

Step 4. *Approximation of the distribution of $\hat{\mu}$.* Sort the bootstrap estimates to obtain $\hat{\mu}_{(1)}^* \leq \hat{\mu}_{(2)}^* \leq \cdots \leq \hat{\mu}_{(B)}^*$.

Step 5. *Confidence interval.* The $100(1 - \alpha)\%$ bootstrap confidence interval is $(\hat{\mu}_{(q_1)}^*, \hat{\mu}_{(q_2)}^*)$, where $q_1 = \lfloor B\alpha/2 \rfloor$ and $q_2 = B - q_1 + 1$. For $\alpha = 0.05$ and $B = 1000$, $q_1 = 25$ and $q_2 = 976$, and the 95% confidence interval is found to be $(6.01, 13.87)$ as compared to the theoretical $(6.85, 13.05)$.

The MATLAB code for the parametric bootstrap procedure of Table 2.5 is given in Section A1.2 of Appendix 1.

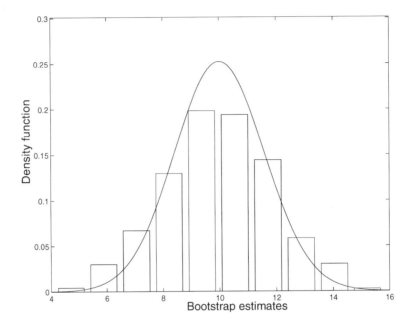

Fig. 2.6. Histogram of $\hat{\mu}_1^*, \hat{\mu}_2^*, \ldots, \hat{\mu}_{1000}^*$, based on the random sample $\mathcal{X} = \{-2.41,$ 4.86, 6.06, 9.11, 10.20, 12.81, 13.17, 14.10, 15.77, 15.79\}, together with the density function of a Gaussian variable with mean 10 and variance 25.

In signal processing and many other engineering applications it is more likely that the type of distribution that generated the data is unknown. Thus, the applicability of the parametric bootstrap is limited. As might be expected, our experience has shown that the use of a non-parametric bootstrap approach is more accurate than a parametric bootstrap procedure with an incorrect model assumption.

2.1.4 Bootstrap resampling for dependent data

Until now we assumed that the random sample $\mathcal{X} = \{X_1, X_2, \ldots, X_n\}$ consists of iid data from a completely unspecified distribution F except for the case of the parametric bootstrap. However, the assumption of iid data can break down in practice either because the data is not independent or because it is not identically distributed. In some practical applications, the data can be dependent and at the same time not identically distributed.

In cases where the underlying model that generated the data is known, we can still invoke the bootstrap principle (Freedman, 1981; Chatterjee, 1986; Bose, 1988; Kreiss and Franke, 1992; Efron and Tibshirani, 1993; Paparoditis, 1996a). For example, assume that the data is identically distributed but not independent such as in autoregressive (AR) models. If no plausible model is available for the probability mechanism generating stationary observations, we could make the assumption of weak dependence.

The assumption of weak dependence is satisfied for processes that fulfil strong mixing conditions.† For processes that contain weakly dependent data the concept of the *moving block* bootstrap has been proposed (Künsch, 1989; Liu and Singh, 1992; Politis and Romano, 1992a, 1994; Bühlmann, 1994). The principle of the moving block bootstrap is as follows. Given n samples, we randomly select blocks of length l from the original data and concatenate them together to form a resample as shown in Figure 2.7, where the block size is set to $l = 20$ and the number of samples is $n = 100$.

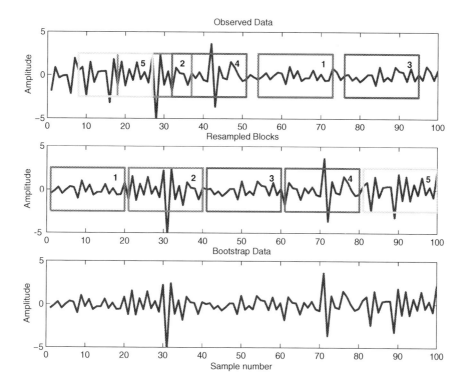

Fig. 2.7. Principle of the moving block bootstrap.

There exist other methods of resampling dependent data, which are closely

† Loosely speaking a process is strong mixing if observations that are far apart in time are almost independent (Rosenblatt, 1985).

related to the moving block bootstrap. For example, the circular blocks bootstrap (Politis and Romano, 1992b; Shao and Yu, 1993) allows blocks which start at the end of the data to wrap around to the start. The block of blocks bootstrap, which is also referred to as the double block bootstrap and originally proposed by Künsch (1989) uses two levels of blocking. The stationary bootstrap is another variation of the moving block bootstrap, where the length of the block is allowed to be random (Politis and Romano, 1994). Other techniques for resampling weakly dependent data involve crossing between the block bootstrap and Richardson extrapolation (Bickel and Yahav, 1988; Hall and Jing, 1994).

Reports on applications of dependent data bootstrap procedures to signal processing include the estimation of confidence intervals for spectra and cross-spectra (Politis *et al.*, 1992b) and estimation of higher order cumulants (Zhang *et al.*, 1993).

We have written two MATLAB functions that, together with bootrsp.m, facilitate bootstrap resampling for weakly dependent data (see Appendix 2). Given the random samples $\mathcal{X} = \{X_1, X_2, \ldots, X_n\}$, the function segments.m obtains Q overlapping $(l > m)$ or non-overlapping $(l \leq m)$ segments each of l samples in the form of a matrix with l rows and q columns, with m being the length of the overlap. The function segmcirc.m, on the other hand, allows the data to be wrapped around in a circle, that is, we define for $i > n - 1$, $X_i = X_{in}$, where $in = i(\text{mod } (n-1))$ and mod denotes the modulo operator. Numerical examples of the moving block bootstrap and the circular block bootstrap are considered next.

The moving block bootstrap creates resamples by concatenating k l-tuples, where $l < n$ and $k = \lceil n/l \rceil$. The l-tuples are drawn from $\hat{F}$, which is an l-dimensional distribution formed by assigning probability mass $1/(n - l + 1)$ to the overlapping blocks

$$
\begin{aligned}
&\{X_1, \ldots, X_l\}, \\
&\quad \{X_2, \ldots, X_{l+1}\}, \\
&\qquad \ddots \\
&\quad\quad \{X_{n-l+1}, \ldots, X_n\}.
\end{aligned}
$$

The choice of l for a small sample size is not straightforward. For the moving block bootstrap to be effective, l should be chosen large enough so that as much of the dependence structure as possible is retained in the overlapping blocks, but not so large that the number of blocks $n - l + 1$ becomes small, resulting in a poor estimate of F. These are conflicting issues and for a fixed sample size n, a choice for l that gives satisfactory results may not be available. Some answers to the choice of the block length exist as it is found in the work of Bühlmann and Künsch (1999). However, applicability of the method to various real-life problems has shown that it does suffer from the choice of the block length.

Example 2.1.7 Moving block bootstrap.

```
>> X=1:10;
>> [X_seg,Q]=segments(X,3,2)

   X_seg =

      1    3    5    7
      2    4    6    8
      3    5    7    9

   Q =

      4

>> ind=bootrsp((1:Q),1)

   ind =

      3    3    2    1

>> X_star=X_seg(:,ind);
>> X_star=X_star(:)';

   X_star =

      5  6  7  5  6  7  3  4  5  1  2  3
```

Similarly, we can use the function segmcirc.m to simulate a circular block bootstrap. Consider the following simple example.

Example 2.1.8 Circular block bootstrap.

```
>> X=1:10;
>> Q=5;
```

```
>> X_seg=segmcirc(X,3,2,Q)

   X_seg =

      1    3    5    7    9
      2    4    6    8   10
      3    5    7    9    1

>> ind=bootrsp((1:Q),1)

   ind =

      5    1    4    4    2

>> X_star=X_seg(:,ind);
>> X_star=X_star(:)'

   X_star =

      9   10    1    1    2    3    7    8    9    7    8    9
      3    4    5
```

Note that the data in Example 2.1.7 are not wrapped and that for the given block and overlap lengths, it is possible to construct only four blocks. Example 2.1.8 indicates that in a circular block bootstrap we can construct any number of blocks. In most bootstrap applications, however, we require the size of the bootstrap resample to match the original sample size.

We can ask the same question as we did earlier for the iid bootstrap resampling. That is, does the bootstrap resample retain the information from the original sample when the data is dependent? The convergence of the bootstrap sample mean described in Result 2.1.1 can be generalised to the case of weakly dependent data.

Result 2.1.4 Bootstrap convergence for the sample mean of m-dependent data.

Let $\theta = \mathsf{E}[X]$, the sample mean $\hat{\theta} = n^{-1} \sum_{i} X_i$ and its bootstrap equivalent

$\hat{\theta}^* = n^{-1} \sum_{i} X_i^*$. *If $\mathsf{E}[|X|^{4+\delta}] < \infty$ for some $\delta > 0$, $l \to \infty$, and $l/n \to 0$,* *then (Liu and Singh, 1992)*

$$\sup_x \left| \mathsf{Prob}_* [\sqrt{kl}(\hat{\theta}^* - \mathsf{E}[\hat{\theta}^*]) \leq x] - \mathsf{Prob}[\sqrt{n}(\hat{\theta} - \theta) \leq x] \right| \to_p 0,$$

where $\to_p$ denotes convergence in probability. Furthermore, if $l/\sqrt{n} \to 0$, then $\mathsf{E}[\hat{\theta}^*]$ above can be replaced by $\hat{\theta}$ (Shao and Tu, 1995, p. 391).

Note that the circular block bootstrap ensures that $\mathsf{E}[\hat{\theta}^*] \equiv \hat{\theta}$ for the sample mean.

2.1.5 Examples of dependent data bootstrap estimation

Below, we will consider several examples of bootstrap resampling where the sampled data are not iid. First we will show how we can invoke the basic bootstrap resampling scheme from Section 2.1 in a case where the model that generated the data is known.

Example 2.1.9 Variance estimation in AR models.

Consider the problem of estimating the variance of the parameter of a first order autoregressive process. One may choose to use dependent data bootstrap techniques to solve this problem. However, knowing the model order we can adapt the independent data bootstrap for this dependent data model.

We collect n observations $x_t, t = 1, \ldots, n$, from the first order autoregressive model

$$X_t + a X_{t-1} = Z_t,$$

where Z_t is stationary white Gaussian noise with $\mathsf{E}[Z_t] = 0$ and autocovariance function $c_{ZZ}(u) = \sigma_Z^2 \delta(u)$, where $\delta(u)$ is Kronecker's delta function, defined by

$$\delta(u) = \begin{cases} 1, & u = 0 \\ 0, & \text{otherwise} \end{cases}$$

and a is a real number, satisfying $|a| < 1$. Note that the Gaussian assumption is unnecessary, but merely chosen to enable comparison with existing results. After centring the data which is equivalent to replacing x_t by

$$x_t - \frac{1}{n} \sum_{t=1}^{n} x_t,$$

we fit the first-order autoregressive model to the observation x_t. With the empirical autocovariance function of x_t,

$$
\hat{c}_{xx}(u) =
\begin{cases}
\dfrac{1}{n} \displaystyle\sum_{t=1}^{n-|u|} x_t\, x_{t+|u|}, & 1 \le |u| \le n \\[2mm]
0, & \text{otherwise,}
\end{cases}
\tag{2.3}
$$

we calculate the maximum likelihood estimate of a,

$$
\hat{a} = -\frac{\hat{c}_{xx}(1)}{\hat{c}_{xx}(0)},
\tag{2.4}
$$

which has approximate variance (Priestley, 1981)

$$
\hat{\sigma}_{\hat{a}}^2 = \frac{1 - a^2}{n}.
\tag{2.5}
$$

Although it is necessary to assume Gaussianity of Z_t for obtaining the result in (2.5) (Priestley, 1981), we note that under some regularity conditions an asymptotic formula for $\hat{\sigma}_{\hat{a}}^2$ can be found in the non-Gaussian case and is a function of a and the variance and kurtosis of Z_t (Porat and Friedlander, 1989; Friedlander and Porat, 1995). Below, in Table 2.6, we show how we can approximate $\hat{\sigma}_{\hat{a}}^2$ without knowledge of the distribution of Z_t.

Consider the following experiment where we chose $a = -0.6$, $n = 128$ and Z_t to be Gaussian. The maximum likelihood estimate (MLE), derived from (2.4), was found to be $\hat{a} = -0.6351$, and the standard deviation $\hat{\sigma}_{\hat{a}} = 0.0707$, when applying (2.5). Using the procedure described in Table 2.6, we obtain the histogram of $B = 1000$ bootstrap estimates of a, $\hat{a}_1^*, \hat{a}_2^*, \ldots, \hat{a}_{1000}^*$, shown in Figure 2.8. We then find an estimate of the standard deviation of $\hat{a}$, $\hat{\sigma}_{\hat{a}}^* = 0.0712$, to be close to the theoretical value $\hat{\sigma}_{\hat{a}}$. For comparison purposes we also show in Figure 2.8 as a solid line the kernel density estimator of $\hat{a}$ based on 1000 Monte Carlo simulations. The estimate of the standard deviation of $\hat{a}$, $\hat{\sigma}_{\hat{a}}$, in this case is found to be 0.0694. A MATLAB code of the above procedure of finding the variance of the estimate of the parameter of the AR(1) process is given in Section A1.3 of Appendix 1.

We emphasise that in the bootstrap procedure neither the assumption of a Gaussian distribution for the noise process Z_t nor knowledge of any characteristic of the non-Gaussian distribution is necessary. We would have found an estimate of the variance of $\hat{a}$ for any distribution of Z_t. However, this is not necessarily possible using analytic means because a formula for $\hat{\sigma}_{\hat{a}}^2$ such as the one given in (2.5) is not applicable in the non-Gaussian case.

Table 2.6. *Bootstrap principle for estimating the variance of the estimate of the parameter of an AR(1) process.*

Step 0. *Experiment.* Conduct the experiment and collect n observations x_t, $t = 1, \ldots, n$, from an autoregressive process of order one, X_t.

Step 1. *Calculation of the residuals.* With the estimate $\hat{a}$ of a, define the residuals $\hat{z}_t = x_t + \hat{a} \cdot x_{t-1}$ for $t = 2, 3, \ldots, n$.

Step 2. *Resampling.* Create a bootstrap sample $x_1^*, x_2^*, \ldots, x_n^*$ by drawing $\hat{z}_2^*, \hat{z}_3^*, \ldots, \hat{z}_n^*$, with replacement, from the residuals $\hat{z}_2, \hat{z}_3, \ldots, \hat{z}_n$, then letting $x_1^* = x_1$, and $x_t^* = -\hat{a} x_{t-1}^* + \hat{z}_t^*$, $t = 2, 3, \ldots, n$.

Step 3. *Calculation of the bootstrap estimate.* After centring the data $x_1^*, x_2^*, \ldots, x_n^*$, obtain $\hat{a}^*$, using formulae (2.4) and (2.5) but based on $x_1^*, x_2^*, \ldots, x_n^*$, rather than $x_1, x_2, \ldots, x_n$.

Step 4. *Repetition.* Repeat Steps 2–3 a large number of times, $B = 1000$, say, to obtain $\hat{a}_1^*, \hat{a}_2^*, \ldots, \hat{a}_B^*$.

Step 5. *Variance estimation.* From $\hat{a}_1^*, \hat{a}_2^*, \ldots, \hat{a}_B^*$, approximate the variance of $\hat{a}$ by

$$\hat{\sigma}_{\hat{a}}^{*2} = \frac{1}{B-1} \sum_{i=1}^{B} \left(\hat{a}_i^* - \frac{1}{B} \sum_{j=1}^{B} \hat{a}_j^* \right)^2 .$$

The only assumption we made is that the random variables $Z_1, Z_2, \ldots, Z_n$ are independent and identically distributed.

Note also that we could have obtained $\hat{\sigma}_{\hat{a}}^*$ using the parametric bootstrap. Herein, we could also sample $\hat{z}_t^*$ from a fitted Gaussian distribution, i.e., a Gaussian distribution with mean zero and variance $\hat{\sigma}_Z^2 = (1 - \hat{a}^2)\hat{c}_{xx}(0)$, instead of resampling from the residuals $\hat{z}_2, \hat{z}_3, \ldots, \hat{z}_n$. The bootstrap estimate $\hat{a}^*$ is then computed from the so obtained resamples. This approach however, would require knowledge of the distribution of Z_t, $t = 1, \ldots, n$. Of course, one may make this assumption in the absence of knowledge of the distribution, but as mentioned earlier, the validity of the results when the assumption is not adequate is not given.

We would have taken a similar approach to estimate the variances and the covariances of the parameter estimates of an AR(p) process, where p is the order of the autoregressive process. For more details on regression analysis using the bootstrap see the work of Politis (1998).

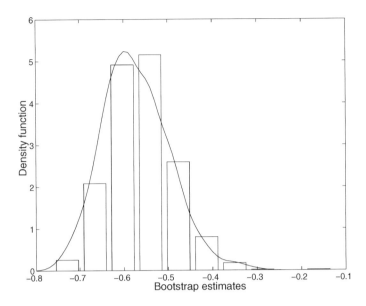

Fig. 2.8. Histogram of $\hat{a}_1^*, \hat{a}_2^*, \ldots, \hat{a}_{1000}^*$ for an AR(1) process with parameter $a = -0.6$, $n = 128$ and Z_t Gaussian. The MLE for a was $\hat{a} = -0.6351$ and $\hat{\sigma}_{\hat{a}} = 0.0707$. The bootstrap estimate was $\hat{\sigma}_{\hat{a}}^* = 0.0712$ as compared to $\hat{\sigma}_{\hat{a}} = 0.0694$ based on 1000 Monte Carlo simulations.

Let us consider again the problem of finding a confidence interval for the power spectral density of a stationary random process that was introduced in Chapter 1. This time, however, we will consider that the observations are taken from a weakly dependent time series.

Example 2.1.10 Confidence interval estimation for the power spectral density: a residual based method.

Let $X_1, \ldots, X_n$ be observations from a strictly stationary real-valued time series X_t with mean zero, finite variance, and spectral density $C_{XX}(\omega)$ as defined in (1.1). The bootstrap can be used in different ways to estimate a confidence interval for $C_{XX}(\omega)$. The first method, proposed by Franke and Härdle (1992), is based on bootstrapping the residuals

$$\varepsilon_k = \frac{I_{XX}(\omega_k)}{C_{XX}(\omega_k)}, \quad k = 1, \ldots, M,$$

where $I_{XX}(\omega)$ is the periodogram of the sample which is defined in (1.2). Following our previous example where we resampled AR residuals, we can devise an algorithm for bootstrapping the spectral density as shown in Figure 2.9 and Table 2.7.

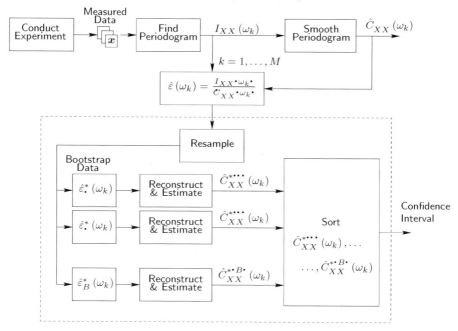

Fig. 2.9. The general concept of the residual based bootstrap confidence interval estimation for the spectral density $C_{XX}(\omega)$.

Suppose now that the real-valued time series X_t is an AR process of order $p = 5$,

$$X_t = 0.5X_{t-1} - 0.6X_{t-2} + 0.3X_{t-3} - 0.4X_{t-4} + 0.2X_{t-5} + \varepsilon_t \,,$$

where ε_t is an iid Gaussian process with zero mean and unit variance. The spectral density of X_t is given by

$$C_{XX}(\omega) = \frac{\sigma^2}{2\pi \left|1 + \displaystyle\sum_{l=1}^{p} a_l \exp(-j\omega l)\right|^2}, \qquad (2.6)$$

where a_l, $l = 1, \ldots, p$ are the coefficients of the AR process, and σ^2 is the variance of the noise. Let $n = 256$ and consider the estimation of $C_{XX}(\omega_k)$ at frequencies $\omega_k = 2\pi k/256$ for

$$k = 41, 42, 43, 66, 67, 68, 83, 84, 85,$$

which correspond to two peaks and the trough between both peaks (see the solid-line plot in Figure 2.10). We run the residual based algorithm of Table 2.7 for several values of $m = \lfloor (h\,n - 1)/2 \rfloor$ to find a suitable global

Table 2.7. *A residual based bootstrap method for estimating a confidence interval for the spectral density.*

Step 1. *Compute residuals.* Choose a $h > 0$ which does not depend on ω and compute

$$\hat{\varepsilon}_k = \frac{I_{XX}(\omega_k)}{\hat{C}_{XX}(\omega_k; h)}, \quad k = 1, \ldots, M.$$

Step 2. *Rescaling.* Rescale the empirical residuals to

$$\tilde{\varepsilon}_k = \frac{\hat{\varepsilon}_k}{\hat{\varepsilon}}, \quad k = 1, \ldots, M$$

where

$$\hat{\varepsilon} = \frac{1}{M} \sum_{j=1}^{M} \hat{\varepsilon}_j.$$

Step 3. *Resampling.* Draw independent bootstrap residuals $\tilde{\varepsilon}_1^*, \ldots, \tilde{\varepsilon}_M^*$ from the empirical distribution of $\tilde{\varepsilon}_1, \ldots, \tilde{\varepsilon}_M$.

Step 4. *Bootstrap estimates.* With a bandwidth g, find

$$I_{XX}^*(\omega_k) = I_{XX}^*(-\omega_k) = \hat{C}_{XX}(\omega_k; g)\, \tilde{\varepsilon}_k^*, \quad k = 1, \ldots, M,$$

$$\hat{C}_{XX}^*(\omega; h) = \frac{1}{n\,h} \sum_{k=-M}^{M} K\left(\frac{\omega - \omega_k}{h}\right) I_{XX}^*(\omega_k).$$

Step 5. *Confidence bands estimation.* Repeat Steps $3-4$ and find c_U^* (and proceed similarly for c_L^*) such that

$$\mathsf{Prob}_* \left[\sqrt{n\,h}\, \frac{\hat{C}_{XX}^*(\omega; h) - \hat{C}_{XX}(\omega; g)}{\hat{C}_{XX}(\omega; g)} \le c_U^* \right] = \frac{\alpha}{2}.$$

That is

$$\frac{\hat{C}_{XX}(\omega; h)}{1 + c_U^*(n\,h)^{-1/2}}$$

is the upper bound of a $100(1 - \alpha)\%$-confidence interval for $C_{XX}(\omega)$.

---

bandwidth (Franke and Härdle, 1992). It lies somewhere around $h = 0.1$ which corresponds to $m = 6$. We then proceed with $B = 1000$ bootstrap replications, calculate the confidence bands for each replication, and derive the coverage percentage of the true spectral density given by (2.6), for both the residual based and the χ^2 based approximations as outlined in Chapter 1.

The results are shown in Table 2.8, where we provide the lower bound, the upper bound, and the coverage for the above frequencies. A typical result showing confidence intervals for spectral density of X_t for one realisation is presented in Figure 2.10 along with the true density given by (2.6).

Table 2.8. *Comparative results of the performance of the residual based bootstrap method and the χ^2 method for an AR(5) process driven by Gaussian noise.*

ω_k	Residual based method			χ^2 method		
	Lower	Upper	Coverage	Lower	Upper	Coverage
1.0063	0.028	0.089·	0.883	0.005	0.083	0.912
1.0308	0.029	0.097	0.874	0.004	0.088	0.908
1.0554	0.027	0.101	0.872	0.004	0.082	0.914
1.6199	0.058	0.065	0.877	0.034	0.009	0.957
1.6444	0.056	0.071	0.873	0.035	0.007	0.958
1.6690	0.054	0.069	0.877	0.030	0.012	0.958
2.0371	0.037	0.096	0.867	0.011	0.074	0.915
2.0617	0.036	0.097	0.867	0.007	0.080	0.913
2.0862	0.034	0.094	0.872	0.008	0.090	0.902

Now, we run the same experiment for a non-Gaussian process,

$$Y_t = Y_{t-1} - 0.7Y_{t-2} - 0.4Y_{t-3} + 0.6Y_{t-4} - 0.5Y_{t-5} + \zeta_t,$$

where ζ_t are independent uniformly distributed variables $\mathcal{U}[-\pi, \pi)$, with mean zero and variance $\sigma^2 = \pi^2/12$. Here again $n = 256$, but we consider the frequencies $\omega_k = 2\pi k/256$ of Y_t for

$$k = 26, 27, 28, 52, 53, 54, 95, 96, 97,$$

which correspond to two peaks and the minimum of the true spectral density (see solid line plot in Figure 2.11). The results for 1000 replications are presented in Table 2.9, where we provide the lower bound, the upper bound, and the coverage. A typical result of confidence bands of the spectral density of Y_t for one realisation is presented in Figure 2.11 along with the true density given by (2.6). A MATLAB function bspecest.m for bootstrapping kernel spectral densities based on resampling from the periodogram of the original data is given in Appendix 2.

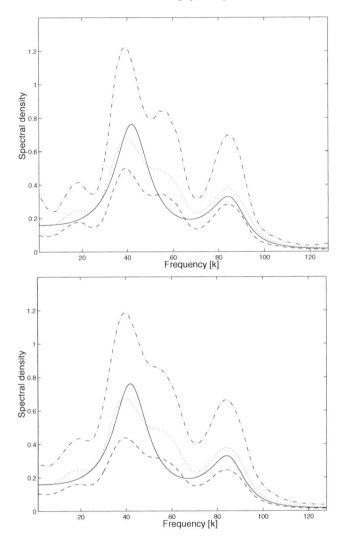

Fig. 2.10. 95% confidence bands for the spectral density of a strictly stationary real-valued time series X_t in Example 2.1.10 using the residual based bootstrap method (top) and the χ^2 approximation (bottom). The true spectral density and its kernel estimate is indicated by a solid and dotted line, respectively.

The results provided in the previous example clearly show that the confidence interval for spectral densities derived using the bootstrap approximates the theoretical result. The advantage of the bootstrap method lies in the fact that it does not assume the distribution of the process to be known or make the assumption that the sample size is large. Thus, it can be used in a more general context of power spectral density estimation.

Table 2.9. *Comparative results of the performance of the residual based bootstrap method and the χ^2 method for an AR(5) process driven by uniformly distributed noise.*

	Residual method			χ^2 method		
ω_k	Lower	Upper	Coverage	Lower	Upper	Coverage
0.6381	0.003	0.200	0.797	0.000	0.247	0.753
0.6627	0.001	0.250	0.749	0.000	0.339	0.661
0.6872	0.001	0.297	0.702	0.000	0.409	0.591
1.2763	0.000	0.451	0.549	0.000	0.635	0.365
1.3008	0.001	0.408	0.591	0.000	0.554	0.446
1.3254	0.002	0.297	0.701	0.001	0.393	0.606
2.3317	0.025	0.108	0.867	0.009	0.028	0.963
2.3562	0.026	0.113	0.861	0.007	0.029	0.964
2.4298	0.028	0.113	0.859	0.007	0.035	0.958

The above examples indicate that in signal processing applications we are often able to reformulate the problem and still use the basic resampling scheme for iid data. Let us consider another example, in which we resample residuals.

Example 2.1.11 Confidence intervals for the bicoherence.

Given observations of a process, we wish to estimate confidence bands for the bicoherence, defined as a variance-normalised magnitude of the bispectrum (Brillinger, 1981). Suppose that $X_1, \ldots, X_n$ are a model for observations from a strictly stationary real-valued random process X_t, $t \in \mathbb{Z}$, with zero mean $\mu_X = 0$, a finite variance $\sigma_X^2 < \infty$, and bispectrum $C_{XXX}(\omega_j, \omega_k)$, $-\pi \le (\omega_j, \omega_k) \le \pi$.

With Cramér representation (Brillinger, 1981)

$$X(t) = \int_{-\pi}^{\pi} e^{j\omega t} dZ_X(\omega),$$

where $dZ_X(\omega)$ are orthogonal increments, the bispectrum is defined through

$$
\begin{aligned}
C_{XXX}(\omega_j, \omega_k) d\omega_j d\omega_k &= \int_{-\pi}^{\pi} \mathsf{E}\left[dZ_X(\omega_j) dZ_X(\omega_k) \bar{dZ}_X(\omega_j + \omega_k)\right] \\
&= \int_{-\pi}^{\pi} \mathrm{cum}\left[dZ_X(\omega_j), dZ_X(\omega_k), dZ_X(-\omega_j - \omega_k)\right],
\end{aligned}
$$

where $\mathrm{cum}[\cdot]$ denotes the cumulant.

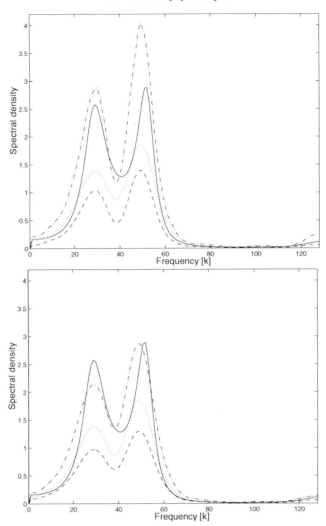

Fig. 2.11. 95% confidence bands for the spectral density of a strictly stationary real-valued time series Y_t in Example 2.1.10 using the residual based bootstrap method (top) and the χ^2 approximation (bottom). The true spectral density and its kernel estimate is indicated by a solid and dotted line respectively.

Let us divide the observations X_t, $t = 0, \ldots, T-1$ into P non-overlapping segments of n consecutive measurements and calculate for each segment $i = 1, \ldots, P$ the biperiodogram defined as

$$I_{XXX}^{(i)}(\omega_j, \omega_k) = \frac{1}{n} d_X^{(i)}(\omega_j) d_X^{(i)}(\omega_k) \bar{d}_X^{(i)}(\omega_j + \omega_k), \quad -\pi \leq \omega_j, \omega_k \leq \pi,$$

with $d_X^{(i)}(\omega_i)$ being the finite Fourier transform of the ith segment and $\bar{d}_X$ its complex conjugate. An estimate of $C_{XXX}(\omega_j, \omega_k)$ is obtained through

$$\hat{C}_{XXX}(\omega_j, \omega_k) = \frac{1}{P} \sum_{i=1}^{P} I_{XXX}^{(i)}(\omega_j, \omega_k).$$

Other estimators of the bispectrum such as the smoothed biperiodogram may be used instead. It is important that the estimator is consistent; knowledge of its (asymptotic) distribution is unnecessary for confidence interval estimation.

An independent data bootstrap based on resamples of $\hat{C}_{XXX}(\omega_j, \omega_k)$ for estimation of its distribution is not straightforward due to the structure of $\hat{C}_{XXX}(\omega_j, \omega_k)$, including dependence of bispectral estimates (Zoubir and Iskander, 1999).

Potentially a block of blocks bootstrap procedure described in Section 2.1.4 can be used to resample the bispectrum estimates. However, the procedure requires much larger sample sizes limiting its application to real-world problems. One may argue, that an approach similar to that of bootstrapping kernel estimates as shown in the previous example can be used (Franke and Härdle, 1992; Chen and Romano, 1999).

However, the way we form the approximate regression is not always straightforward. Here, we write

$$I_{XXX}^{(i)}(\omega_j, \omega_k) = C_{XXX}(\omega_j, \omega_k) + \varepsilon_{j,k} V(\omega_j, \omega_k), \qquad (j, k) \in \mathcal{D}, \qquad (2.7)$$

where

$$V(\omega_j, \omega_k)^2 = n C_{XX}(\omega_j) C_{XX}(\omega_k) C_{XX}(\omega_j + \omega_k)$$
$$\times [1 + \delta(j - k) + \delta(n - 2j - k) + 4\delta(n - 3j)\delta(n - 3k)],$$

$C_{XX}(\omega)$ is the spectrum of X_t, $\delta(k)$ is Kronecker's delta function, $\omega_j = 2\pi j/n$ and $\omega_k = 2\pi k/n$ are discrete frequencies and $\mathcal{D} = \{0 < k \le j, 2j + k \le n\}$. From the above regression, we can design a procedure similar to the one we used for estimating confidence intervals for power spectra.

To resample the residuals in the regression (2.7) we assume that $\varepsilon_{j,k}$ are independent and identically distributed random variates. This assumption holds for a reasonably large n. We note that the magnitude of the correlation between biperiodograms $I_{XXX}^{(i)}(\omega_j, \omega_k)$ and $I_{XXX}^{(i)}(\omega_{j'}, \omega_{k'})$ is approximately of order $O(n^{-2})$ for $j \ne j'$ and $k \ne k'$ ($O(n^{-1})$ for "exclusive or" cases). The complete bootstrap procedure is shown in Table 2.10.

Table 2.10. *A bootstrap algorithm for estimating a confidence interval for the bicoherence.*

Step 1. *Test statistics:* Calculate $I_{XX}^{(i)}(\omega_j)$, $I_{XXX}^{(i)}(\omega_j, \omega_k)$, $\hat{C}_{XX}(\omega_j)$, $\hat{C}_{XXX}(\omega_j, \omega_k)$, $\hat{\sigma}(\omega_j, \omega_k)$ (using the bootstrap) and

$$\tilde{C} = \sum_{j,k \in \mathcal{D}} \frac{|\hat{C}_{XXX}(\omega_j, \omega_k)|}{\hat{\sigma}(\omega_j, \omega_k)}.$$

Step 2. *Calculate residuals:* For each segment, estimate the residuals

$$\hat{\varepsilon}_{j,k}^{(i)} = \frac{I_{XXX}^{(i)}(\omega_j, \omega_k) - \hat{C}_{XXX}(\omega_j, \omega_k)}{\hat{V}(\omega_j, \omega_k)}, \qquad j, k \in \mathcal{D}.$$

Step 3. *Centre residuals:* Centre the residuals to obtain $\tilde{\varepsilon}_{j,k}^{(i)} = \hat{\varepsilon}_{j,k}^{(i)} - \bar{\varepsilon}^{(i)}$, $i = 1, \ldots, P$, where $\bar{\varepsilon}^{(i)}$ is an average over all $\hat{\varepsilon}_{j,k}^{(i)}$.

Step 4. *Resampling:* Draw independent bootstrap residuals $\tilde{\varepsilon}_{j,k}^{(i)*}$.

Step 5. *Compute bootstrap estimates:* Compute the bootstrap biperiodogram

$$I_{XXX}^{(i)*}(\omega_j, \omega_k) = \hat{C}_{XXX}(\omega_j, \omega_k) + \tilde{\varepsilon}_{j,k}^{(i)*} \hat{V}(\omega_j, \omega_k),$$

obtain the bootstrap bispectral estimate

$$\hat{C}_{XXX}^{*}(\omega_j, \omega_k) = \frac{1}{P} \sum_{i=1}^{P} I_{XXX}^{(i)*}(\omega_j, \omega_k),$$

and compute the bootstrap estimate of bicoherence

$$\tilde{C}^{*} = \frac{\hat{C}_{XXX}^{*}(\omega_j, \omega_k)}{\hat{\sigma}(\omega_j, \omega_k)}.$$

Step 6. *Repetition:* Repeat Steps 4 and 5 a large number of times, to obtain a total of B bootstrap estimates $\tilde{C}_1^{*}, \ldots, \tilde{C}_B^{*}$.

Step 7. *Confidence band estimation:* Rank the collection $\tilde{C}_1^{*}, \ldots, \tilde{C}_B^{*}$ into increasing order to obtain $\tilde{C}_{(1)}^{*} \leq \cdots \leq \tilde{C}_{(B)}^{*}$, for each frequency pair (ω_j, ω_k). The desired $100(1 - \alpha)\%$ bootstrap confidence interval is $(\tilde{C}_{(q_1)}^{*}, \tilde{C}_{(q_2)}^{*})$, where $q_1 = \lfloor B\alpha/2 \rfloor$ and $q_2 = B - q_1 + 1$.

In Figure 2.12, we show the lower and upper 95% confidence band for the bicoherence of a white Gaussian process. In this example the total sample size is $T = 4096$ while the segment length is $n = 64$.

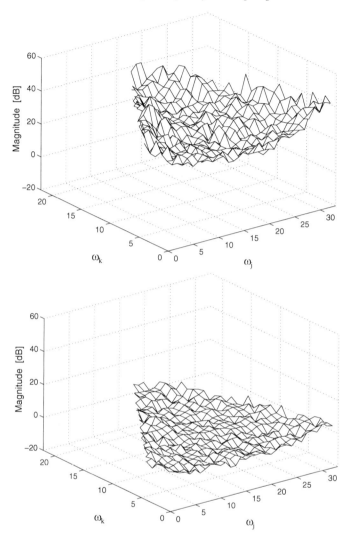

Fig. 2.12. Lower (bottom) and upper (top) confidence bands of the bicoherence of a white Gaussian process in Example 2.1.11.

In the following example, we continue with confidence interval estimation for spectral densities, but this time we will use the block of blocks (double block) and the circular block bootstrap methods. The block of blocks bootstrap method was proposed for setting confidence bands for spectra by Politis *et al.* (1992b) and its principle is described in Table 2.11.

Example 2.1.12 Confidence interval estimation for spectral densities: the block of blocks bootstrap method.

Consider the two AR processes from Example 2.1.10

$$X_t = 0.5X_{t-1} - 0.6X_{t-2} + 0.3X_{t-3} - 0.4X_{t-4} + 0.2X_{t-5} + \varepsilon_t,$$

and

$$Y_t = Y_{t-1} - 0.7Y_{t-2} - 0.4Y_{t-3} + 0.6Y_{t-4} - 0.5Y_{t-5} + \zeta_t.$$

We will now use a larger sample with $n = 2000$ so that it is feasible to perform double block bootstrapping. The block sizes and overlaps are set to $L = 128$, $M = 20$, $l = 6$, and $h = 2$. The number of bootstrap resamples is chosen to be $B = 100$. Typical results of confidence bands of the spectral density of X_t and Y_t are shown in Figure 2.13. The true densities are indicated by solid lines.

Table 2.11. *Block of blocks bootstrap for estimating a confidence interval for spectral densities.*

Step 1. *First block.* Given $X_1, \ldots, X_n$, obtain Q overlapping $(0 < M < L)$ or non-overlapping $(M = 0)$ segments of L samples and estimate $\hat{C}_{XX}^{(i)}(\omega)$, $i = 1, \ldots, Q$.

Step 2. *Second block.* Divide $\hat{C}_{XX}^{(1)}(\omega), \ldots, \hat{C}_{XX}^{(Q)}(\omega)$ into q overlapping $(0 < h < l)$ or non-overlapping $(h = 0)$ blocks, say $\mathcal{C}_j$, $j = 1, \ldots, q$, each containing l estimates.

Step 3. *Resampling.* Generate k bootstrap samples $y_1^*, \ldots, y_k^*$ of size l each, from $\mathcal{C}_1, \ldots, \mathcal{C}_q$.

Step 4. *Reshaping.* Concatenate $y_1^*, \ldots, y_k^*$ into a vector $\mathbf{Y}^*$ and estimate $\hat{C}_{XX}^*(\omega)$.

Step 5. *Confidence interval.* Repeat Steps 3–4 and proceed as before to obtain a confidence interval for $C_{XX}(\omega)$.

In the case where one has n observations, the number of blocks that can be formed is given by $Q = \lfloor (n-L)/M \rfloor + 1$, where $L - M$ denotes the overlapping distance. Thus, there are situations where not all information is retrieved from the data and the distribution of an estimate derived from blocks is biased. A simple way to have an unbiased bootstrap distribution is to wrap

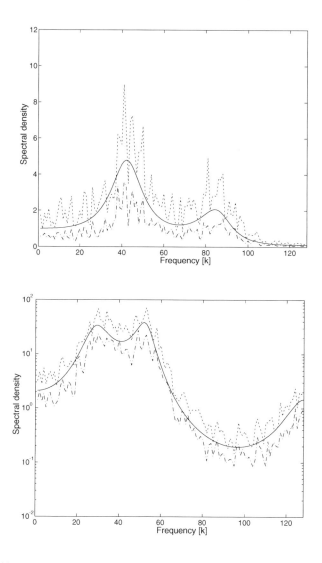

Fig. 2.13. 95% confidence bands for the spectral densities of X_t (top) and Y_t (bottom) from Example 2.1.12 using the block of blocks bootstrap method. The true spectral densities are indicated by solid lines.

the observations X_i around the circle and use the circular block bootstrap (Politis and Romano, 1992b). The advantage is to be able to set the number of blocks Q or q so as to obtain unbiased estimates. Typical results of confidence bands for the spectral densities of the considered AR processes X_t and Y_t are shown in Figure 2.14 when using the circular block bootstrap. Again, the true densities are indicated by solid lines. For $n = 2000$ there is not much difference between the two results in Figures 2.13 and 2.14. The difference between the block of blocks and the circular bootstrap will be more pronounced when significant data is left in the moving block resampling. A MATLAB function bspecest2.m for bootstrapping spectral densities based on a moving block bootstrap is given in Appendix 2.

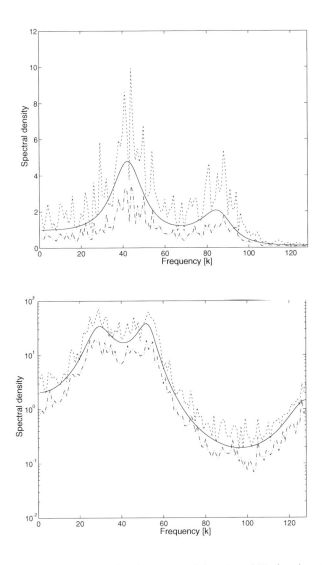

Fig. 2.14. 95% confidence bands for the spectral density of X_t (top) and Y_t (bottom) from Example 2.1.12 using the circular block bootstrap. The true spectral densities are indicated by solid lines.

A rigorous performance analysis of the block of blocks and circular bootstrap procedures as compared to the χ^2 method is difficult because there are many parameters to be optimised such as the length of the segments in each block and the length of the overlapping distances in the block of blocks bootstrap or the number of segments in each block in the circular block bootstrap. The reader interested in these issues is referred to the work of Politis and Romano (1992a,b, 1994) and Politis *et al.* (1992b). Other applications of bootstrap methods in the context of spectral density estimation can be found in the work of Paparoditis (1996b); Paparoditis and Politis (1999); Zoubir and Iskander (1999). An interesting more recent technique of resampling weakly dependent data is called the threshold bootstrap (Park and Willemain, 1999). In this method a threshold is set across the data and a cycle is defined that consists of alternating runs above and below the threshold. The data is resampled in blocks of a length that is some multiple of the cycle.

2.2 The principle of pivoting and variance stabilisation

We mentioned in an earlier example (see Table 2.4) that a straightforward application of the bootstrap for confidence interval estimation leads to a coverage that is smaller than its nominal value. One way to remedy this problem is to consider a pivotal statistic. A statistic $T_n(X, \theta)$ is called pivotal if it possesses a fixed probability distribution independent of θ (Lehmann, 1991; Cramér, 1999).

Hall (1992) notes that bootstrap confidence intervals or tests have excellent properties even for a relatively low fixed number of resamples. For example, one can show that the coverage error in confidence interval estimation with the bootstrap is of order $O_p(n^{-1})$ as compared to $O_p(n^{-1/2})$ when using the Gaussian approximation. The claimed accuracy holds whenever the statistic is asymptotically pivotal (Hall and Titterington, 1989; Hall, 1992). A confidence interval based on a pivotal bootstrap statistic is known as a *percentile-t confidence interval* (Efron, 1987; Hall, 1988).

To ensure pivoting, the statistic is usually "studentised", i.e., we form

$$T_n = \frac{\hat{\theta} - \theta}{\hat{\sigma}_{\hat{\theta}}}.$$

Using the above transformation for location statistics, such as the sample mean and the sample median leads to good results with the bootstrap (Efron and Tibshirani, 1993). Note, however, that pivoting often does not hold unless a variance stabilising transformation for the parameter estimate of interest is applied first.

There are situations where it is possible to find a variance stabilising transformation such as in the case of the correlation coefficient as we will discuss in an example. However, in many applications a variance stabilising transformation for a statistic of interest is not known. An interesting approach for estimating variance stabilising transformations was introduced by Tibshirani (1988). The method estimates a variance stabilising transformation using an initial double bootstrap step, computes a so-called percentile-*t* confidence interval for the transformed statistic, and then transforms this interval back to the original scale. Tibshirani's idea can be used to construct confidence intervals for parameter estimators or statistical tests as it will be shown in Chapter 3. Advantages of Tibshirani's approach is that it is "automatic"– the stabilising transformation is derived from the data and does not need to be known in advance, and that it is invariant under monotonic transformations. The bootstrap principle for variance stabilisation along with bootstrap quantile estimation is outlined in Table 2.12.

Table 2.12. *The principle of variance stabilisation and quantile estimation.*

Step 1. *Estimation of the variance stabilising transformation.*

    (a) Generate B_1 bootstrap samples $\mathcal{X}_i^*$ from $\mathcal{X}$ and for each calculate $\ddot{\theta}_i^*$, $i = 1, \ldots, B_1$. Let, for example, $B_1 = 100$.

    (b) Generate B_2 bootstrap samples from $\mathcal{X}_i^*$, $i = 1, \ldots, B_1$, and calculate $\hat{\sigma}_i^{*2}$, a bootstrap estimate for the variance of $\hat{\theta}_i^*$, $i = 1, \ldots, B_1$. Let, for example, $B_2 = 25$.

    (c) Estimate the variance function $\zeta(\theta)$ by smoothing the values of $\hat{\sigma}_i^{*2}$ against $\hat{\theta}_i^*$, using, for example, a fixed-span 50% "running lines" smoother.

    (d) Estimate the variance stabilising transformation $h(\hat{\theta})$ from

$$h(\theta) = \int^{\theta} \{\zeta(s)\}^{-1/2} ds.$$

Step 2. *Bootstrap quantile estimation.* Generate B_3 bootstrap samples and compute $\hat{\theta}_i^*$ and $h(\hat{\theta}_i^*)$ for each sample i. Approximate the distribution of $h(\hat{\theta}) - h(\theta)$ by that of $h(\hat{\theta}^*) - h(\hat{\theta})$.

Section A1.4 of Appendix 1 provides a MATLAB routine that performs variance stabilisation. We use the sample mean as our statistic $\hat{\theta}$. The routine uses the function smooth.m which is a running line smoother that fits the data by linear least squares (Hastie and Tibshirani, 1990) (see Appendix 2). This smoother fits a least-squares regression line in symmetric windows centred at each $\hat{\theta}_i^*$, $i = 1, \ldots, B_1$. Note that any other smoothing procedure is acceptable. We have found that the running line smoother is simple and in most cases performs satisfactorily.

2.2.1 Some examples

In this section, we provide some examples that show how important variance stabilisation is in the estimation of confidence intervals.

Example 2.2.1 Confidence interval for the mean with a pivotal statistic.

Consider again the problem of finding a confidence interval for the mean as in Example 2.1.5. Let $\mathcal{X} = \{X_1, \ldots, X_n\}$ be a random sample from some unknown distribution with mean μ_X and variance σ_X^2. We want to find an estimator of μ_X with a $100(1 - \alpha)\%$ confidence interval. Let $\hat{\mu}_X$ and $\hat{\sigma}_X^2$ be the sample mean and the sample variance of $\mathcal{X}$, respectively. As an alternative to Example 2.1.5, we will base our method for finding a confidence interval for μ_X on the statistic

$$\hat{\mu}_Y = \frac{\hat{\mu}_X - \mu_X}{\hat{\sigma}}, \tag{2.8}$$

where $\hat{\sigma}$ is the standard deviation of $\hat{\mu}_X$. It is known that the distribution of this statistic is, asymptotically for large n, free of unknown parameters (Kendall and Stuart, 1967). The bootstrap algorithm for estimating a confidence interval for the mean based on the pivotal statistic (2.8) is shown in Table 2.13.

The nested bootstrap for estimating the standard deviation of $\hat{\mu}_X$ in Step 1 of Table 2.13 is performed similarly to the variance estimation presented in Table 2.3 and is as follows. We use a small number of resamples, B_1, where for each resample we estimate μ_X to get $\hat{\mu}_X^{*(b)}$, $b = 1, \ldots, B_1$. Then, we estimate $\hat{\sigma}$ by taking

$$\hat{\sigma}^2 = \frac{1}{B_1 - 1} \sum_{i=1}^{B_1} \left(\hat{\mu}_X^{*(i)} - \sum_{j=1}^{B_1} \hat{\mu}_X^{*(j)} \right)^2.$$

As noted earlier, the number of resamples for variance estimation does not have to be large and is typically set to $B_1 = 25$. The nested bootstrap, often

Table 2.13. *The bootstrap principle for estimating a confidence interval for the mean using a pivotal statistic.*

Step 0. *Experiment.* Conduct the experiment and collect the random data into the sample $\mathcal{X} = \{X_1, X_2, \ldots, X_n\}$.

Step 1. *Parameter estimation.* Based on $\mathcal{X}$, calculate $\hat{\mu}_X$ and its standard deviation $\hat{\sigma}$, using a nested bootstrap.

Step 2. *Resampling.* Draw a random sample, $\mathcal{X}^*$, with replacement, from $\mathcal{X}$.

Step 3. *Calculation of the pivotal statistic.* Calculate the mean of all values in $\mathcal{X}^*$ and using a nested bootstrap, calculate $\hat{\sigma}^*$. Then, form

$$\hat{\mu}_Y^* = \frac{\hat{\mu}_X^* - \hat{\mu}_X}{\hat{\sigma}^*}$$

Step 4. *Repetition.* Repeat Steps 2–3 many times to obtain a total of B bootstrap estimates $\hat{\mu}_{Y,1}^*, \ldots, \hat{\mu}_{Y,B}^*$.

Step 5. *Ranking.* Sort the bootstrap estimates to obtain $\hat{\mu}_{Y,(1)}^* \leq \hat{\mu}_{Y,(2)}^* \leq \cdots \leq \hat{\mu}_{Y,(B)}^*$.

Step 6. *Confidence interval.* If $\left(\hat{\mu}_{Y,(q_1)}^*, \hat{\mu}_{Y,(q_2)}^*\right)$ is an interval containing $(1-\alpha)B$ of the means $\hat{\mu}_Y^*$, where $q_1 = \lfloor B\alpha/2 \rfloor$ and $q_2 = B - q_1 + 1$, then

$$\left(\hat{\mu}_X - \hat{\sigma}\hat{\mu}_{Y,(q_2)}^*, \hat{\mu}_X - \hat{\sigma}\hat{\mu}_{Y,(q_1)}^*\right)$$

is a $100(1 - \alpha)\%$ confidence interval for μ_X.

---

referred to as the double bootstrap, adds a significant computational cost. For example, for typical values of $B = 1000$ and $B_1 = 25$, the total number of bootstrap resamples becomes 25,000. This is not a problem when calculating simple statistics such as the mean. However, in some applications the computational expense may prohibit the use of a bootstrap pivotal statistic. To overcome this problem one may use a variance stabilising transformation as discussed above. For example, if we use the values for $B_1 = 100$, $B_2 = 25$ and $B = 1000$ in Table 2.12, the total number of resamples is reduced from 25,000 to 3,500. More techniques available to reduce the number of computations in a double bootstrap setting are found in the works of Efron (1990); Karlsson and Löthgren (2000).

MATLAB code for estimating the confidence interval for the mean based on a pivotal statistic is shown in Section A1.4 of Appendix 1, where the set of data is the same as in Example 2.1.5.

For the same random sample $\mathcal{X}$ as in Example 2.1.5, we obtained the confidence interval $(3.54, 13.94)$ as compared to the true interval $(6.01, 13.87)$. This interval is larger than the one obtained earlier and enforces the statement that the interval obtained with a non-pivotal statistic has coverage less than the nominal 95%. It also yields better results than an interval derived using the assumption that $\hat{\mu}_Y$ is Gaussian distributed or the (better) approximation that $\hat{\mu}_Y$ is t_{n-1} distributed. The interval obtained here accounts for skewness in the underlying population and other errors (Hall, 1988, 1992; Efron and Tibshirani, 1993).

Example 2.2.2 Confidence interval for the correlation coefficient using variance stabilisation.

Let $\theta = \varrho$ be the correlation coefficient of two unknown populations, and let $\hat{\varrho}$ and $\hat{\sigma}^2$ be estimates of ϱ and the variance of $\hat{\varrho}$, respectively, based on $\mathcal{X} = \{X_1, \ldots, X_n\}$ and $\mathcal{Y} = \{Y_1, \ldots, Y_n\}$. Let then $\mathcal{X}^*$ and $\mathcal{Y}^*$ be resamples, drawn with replacement from $\mathcal{X}$ and $\mathcal{Y}$, respectively, and let $\hat{\varrho}^*$ and $\hat{\sigma}^{*2}$ be bootstrap versions of $\hat{\varrho}$ and $\hat{\sigma}^2$.

By repeated resampling from $\mathcal{X}$ and $\mathcal{Y}$ we compute $\hat{s}_\alpha$ and $\hat{t}_\alpha$, such that with $0 < \alpha < 1$

$$\mathsf{Prob}\left[(\hat{\varrho}^* - \hat{\varrho})/\hat{\sigma}^* \leq \hat{s}_\alpha \mid \mathcal{X}, \mathcal{Y}\right] = \frac{\alpha}{2} = \mathsf{Prob}\left[(\hat{\varrho}^* - \hat{\varrho})/\hat{\sigma}^* \geq \hat{t}_\alpha \mid \mathcal{X}, \mathcal{Y}\right].$$

Using the percentile-t method, we can calculate the confidence interval for ϱ using

$$\mathcal{I}(\mathcal{X}, \mathcal{Y}) = \left(\hat{\varrho} + \hat{\sigma}\hat{t}_\alpha, \ \hat{\varrho} + \hat{\sigma}\hat{s}_\alpha\right).$$

For the correlation coefficient a stabilising and normalising transformation is known (Fisher, 1970; Anderson, 1984). It is given by

$$\breve{\varrho} = \tanh^{-1}\hat{\varrho} = \frac{1}{2}\log\frac{1 + \hat{\varrho}}{1 - \hat{\varrho}}.$$

We could first find a confidence interval for $\xi = \tanh^{-1}\varrho$ and then transform the endpoints back with the inverse transformation $\varrho = \tanh \xi$ to obtain a confidence interval for ϱ.

For X and Y bivariate Gaussian ($\breve{\varrho} \sim \mathcal{N}(\xi, 1/(n-3))$), for example, a 95% confidence interval for ϱ is obtained from

$$\left(\tanh\left(\frac{-1.96}{\sqrt{n-3}} + \breve{\varrho}\right), \tanh\left(\frac{1.96}{\sqrt{n-3}} + \breve{\varrho}\right)\right). \qquad (2.9)$$

Suppose that $X = Z_1 + W$ and $Y = Z_2 + W$, where Z_1, Z_2 and W are pairwise independent and identically distributed. One can easily show that the correlation coefficient of X and Y is $\varrho = 0.5$. Let $z_{1,i}$, $z_{2,i}$ and w_i, $i = 1, \ldots, 15$, be realisations of Z_1, Z_2 and W, respectively. Assume further that these realisations are drawn from the Gaussian distribution and calculate x_i, y_i, $i = 1, \ldots, 15$. Based on x_i and y_i, $i = 1, \ldots, 15$, we found $\hat{\varrho} = 0.36$ and using (2.9) the 95% confidence interval $(-0.18, 0.74)$ for ϱ. On the other hand, we used the bootstrap percentile-t method similar to the one described in Table 2.13 for the mean and found with $B = 1000$ the 95% confidence interval to be $(-0.05, 1.44)$. We also used Fisher's z-transform and calculated, based on the bootstrap, a confidence interval for the transformed parameter $\xi = \tanh^{-1}\varrho$, but without assuming a bivariate Gaussian distribution for (X, Y). We then transformed the end points of the confidence interval back with the inverse transformation $\tanh$ to yield the confidence interval $(-0.28, 0.93)$ for ϱ.

In both bootstrap approaches we have used a jackknife variance estimate (Miller, 1974; Hanna, 1989). The jackknife which can be thought of as a resampling procedure without replacement but of size $n - 1$ is explained in Section 3.4. However, any other method for variance estimation is applicable. It should be mentioned that because ϱ is bounded within the interval $[-1, 1]$, the interval found using the percentile-t method is over-covering, being larger than the two other ones. Therefore, finding a confidence interval for the transformed parameter and then transforming the endpoints back with the inverse transformation yields a better interval than the one obtained using the percentile-t method. We have never observed in the simulations we ran, that the transformed percentile-t confidence interval obtained contained values outside the interval $[-1, 1]$.

Furthermore, we considered a bootstrap-based variance stabilising transformation as an alternative to Fisher's z-transform, as illustrated in Table 2.12. The smoothing in Step 1(c) was performed using a fixed-span 50% "running lines" smoother (Hastie and Tibshirani (1990)). The integration in Step 1(d) was approximated by a trapezoid rule. We used $B_1 = 100$, $B_3 = 1000$ and a bootstrap variance estimate with $B_2 = 25$. The so obtained variance stabilising transformation is depicted in Figure 2.15 along with Fisher's z-transform applied to the same bootstrap values $\hat{\theta}_i^*$,

$i = 1, \ldots, 1000$. Note that the scale difference between the two variance stabilising transformations is not of importance. Rather, the particular shape of each of the transformations is responsible for the final estimate.

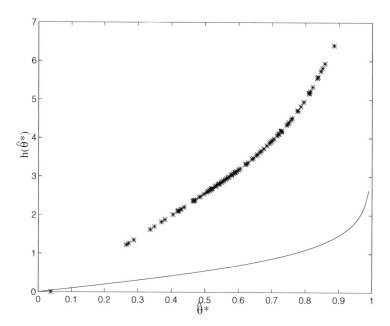

Fig. 2.15. Variance stabilising transformation for the correlation coefficient from Example 2.2.2, estimated using $B_1 = 100$ and $B_2 = 25$. The solid line is a plot of Fisher's z-transform.

To demonstrate the effect of the variance stabilising transformation, we estimated the standard deviation of 1000 bootstrap estimates of $\theta = \varrho$ using the bootstrap (see Table 2.3 with $B = 25$) resulting in the graph of Figure 2.16. The graph shows the dependence of the standard deviation with respect to $\hat{\theta}^* = \hat{\varrho}^*$. After taking 1000 new bootstrap estimates $\hat{\varrho}_i^*$, $i = 1, \ldots, 1000$ and applying the transformation of Figure 2.15, we obtained the more stable standard deviations of Figure 2.17.

For comparison, we have also reproduced in Figure 2.18 the standard deviation of new 1000 bootstrap estimates $\hat{\varrho}_i^*$, $i = 1, \ldots, 1000$ after applying Fisher's z-transform depicted in Figure 2.15 (solid line). The results show that the bootstrap method is satisfactory for estimating the variance stabilising transformation, which will lead to more accurate confidence intervals in situations where a variance stabilising transformation is not known.

To construct a confidence interval for θ with the method of Table 2.12, we first need to find an interval for $h(\theta)$ and then back-transform the interval for $h(\theta)$, to give $(h^{-1}(h(\hat{\theta}) - \hat{t}_{1-\alpha}), h^{-1}(h(\hat{\theta}) - \hat{t}_{\alpha}))$, where $\hat{t}_{\alpha}$ is the αth critical

The bootstrap principle

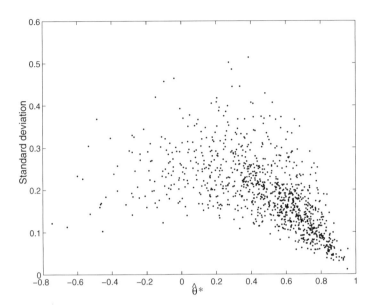

Fig. 2.16. Bootstrap estimates of the standard deviation of $B_3 = 1000$ bootstrap estimates of the correlation coefficient from Example 2.2.2 before variance stabilisation.

point of the bootstrap distribution of $h(\hat{\theta}^*) - h(\hat{\theta})$. For the same data $z_{1,i}$, $z_{2,i}$ and w_i, $i = 1, \ldots, 15$, we obtained the confidence interval $(0.06, 0.97)$. This interval is much tighter than the one obtained using the transformed percentile-t method based on Fisher's z-transform.

As discussed earlier, estimating the variance with a variance stabilising transformation requires much fewer resamples than the equivalent double bootstrap implementation. We will later discuss this issue in the context of hypothesis testing (see Example 3.4.2).

A signal processing application of a bootstrap based variance stabilisation has been reported by Zoubir and Böhme (1995). They considered the problem of sensor placement for knock detection in spark-ignition engines. The approach was to test the closeness to zero of the coherence gain between vibration and cylinder pressure signals. However, the distribution and the variance stabilisation transformation of the test statistic were intractable. In Chapter 5, we summarise the variance stabilisation aspect of this application.

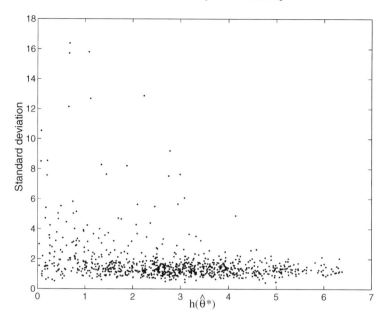

Fig. 2.17. Bootstrap estimates of the standard deviation of $B_3 = 1000$ new bootstrap estimates of the correlation coefficient from Example 2.2.2 after variance stabilisation, obtained through bootstrap. The confidence interval found was $(0.06, 0.97)$.

2.3 Limitations of the bootstrap

The bootstrap does not always work. Examples of bootstrap failure have been reported from the very beginning of its development (Bickel and Freedman, 1981; Mammen, 1992; Young, 1994; Hu and Hu, 2000). Generally speaking, one cannot resample data which stems from a distribution with infinite variance. However, there is still some disagreement on this issue between statisticians (see work of Gine and Zinn (1989) versus that of Athreya (1987)).

Leaving this statistical polemic aside, from an engineering point of view we can assume that the bootstrap principle may not properly work with data that have an infinite variance, for example for α-stable ($\alpha < 2$) distributed data. As noted by Politis (1998), if we take a random sample $\mathcal{X} = \{X_1, \ldots, X_n\}$ from a standard Cauchy distribution, the bootstrap will behave erratically even for a large sample size. Other cases of failure are possible – below we give a simple example.

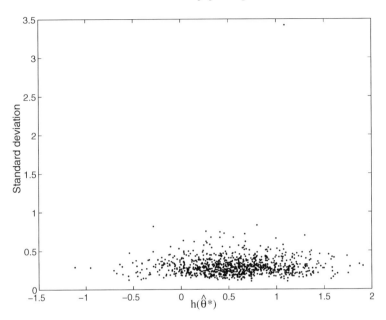

Fig. 2.18. Bootstrap estimates of the standard deviation of $B_3 = 1000$ bootstrap estimates of the correlation coefficient from Example 2.2.2 after applying Fisher's variance stabilising transformation $\tanh^{-1}$.

Example 2.3.1 A failure of the non-parametric bootstrap.

Let $X \sim \mathcal{U}(0, \theta)$ and $\mathcal{X} = \{X_1, X_2, \ldots, X_n\}$. Suppose we wish to estimate θ by $\hat{\theta}$ and its distribution $\hat{F}_{\hat{\Theta}}(\hat{\theta})$. The maximum likelihood estimator of θ is given by

$$\hat{\theta} = \max(X_1, X_2, \ldots, X_n) = X_{(n)}.$$

To obtain an estimate of the density function of $\hat{\theta}$ we sample with replacement from the data and each time estimate $\hat{\theta}^*$ from $\mathcal{X}^*$. Alternatively, we could sample from $\mathcal{U}(0, \hat{\theta})$ and estimate $\hat{\theta}^*$ from $\mathcal{X}^*$ using a parametric bootstrap approach.

  Here, we run an example with $\theta = 1$, $n = 50$ and $B = 1000$. The maximum likelihood estimate of θ is found to be $\hat{\theta} = 0.9843$. In Figure 2.19, we show histograms of $\hat{\theta}^*$ for the non-parametric and the parametric bootstrap. Note that the non-parametric bootstrap shows approximately 64% of values of $\hat{\theta}^*$ that are equal to $\hat{\theta}$. In fact, $\mathsf{Prob}[\hat{\theta}^* = \hat{\theta}] = 1 - (1 - 1/n)^n \longrightarrow 1 - e^{-1} \approx 0.632$ as $n \longrightarrow \infty$. In this example the non-parametric bootstrap failed to capture the distribution of $\hat{\theta}$ while the parametric bootstrap is able to. MATLAB code for this example is given in Section A1.5 of Appendix 1.

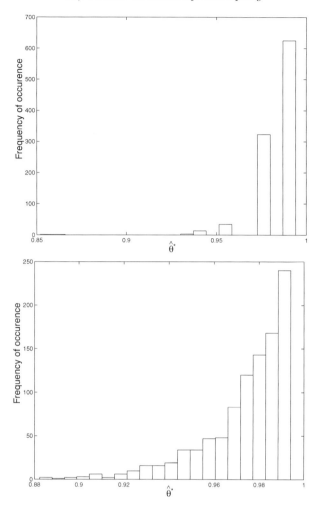

Fig. 2.19. The histogram of $\hat{\theta}^*$ from Example 2.3.1 for a parametric bootstrap (top) and a non-parametric bootstrap (bottom).

2.4 Trends in bootstrap resampling

The concept of bootstrap resampling has been well established over the last two decades. As indicated in the introduction, the bootstrap has found a wide range of applications supporting and in some cases replacing traditional techniques which are usually based on asymptotic approximations. This does not mean that the development of the bootstrap theory is complete.

There are several interesting derivatives of the classical non-parametric boot-strap.

The balanced bootstrap has been suggested by Davison *et al.* (1986). In a balanced bootstrap, also known as Latin hypercube resampling (Hall, 2001), the data is resampled in such a way that each of the data points appears the same number of times in the collection of resamples. This reduces the variability of bootstrap resampling, ensuring that the total mean of bootstrap resamples equals the original sample mean.

An alternative to the balanced bootstrap has been proposed by Rao *et al.* (1997). They note that a bootstrap resample is not equally informative as the original sample, and they propose to resample sequentially with replacement until all data in the original sample is used. This leads to resamples of random lengths.

Several methods have been proposed to minimise the computational cost of bootstrap resampling (Efron, 1990; Karlsson and Löthgren, 2000) and efficient parallel implementation of the bootstrap on a hypercube computer was proposed by Xu and Shiue (1991).

2.5 Summary

In this chapter, we have introduced the ideas behind the bootstrap principle. We have discussed the non-parametric as well as the parametric bootstrap. The exposure of the bootstrap has been supported with numerous examples for both iid as well as dependent data. We have covered bootstrap methods for dependent data, such as the moving block bootstrap, but we have also shown, by way of example, how one can use the independent data bootstrap to tackle complicated data structures such as autoregressions. The examples presented in this chapter demonstrate the power of the bootstrap and its superior performance compared to classical methods based, for example, on asymptotic theory. This being said, care is required when applying the bootstrap. Thus, we reported a case where the bootstrap fails.

The reader should note the following points, which we have emphasised throughout the chapter.

- The parametric bootstrap outperforms the non-parametric bootstrap when the number of samples is small. However, this holds only if the assumed model is correct. Generally speaking, the non-parametric bootstrap is more robust than the parametric bootstrap.
- It is relatively easy to use the bootstrap in linear models if the errors are iid. This, however, may not be given in practice. The approach would

be to use a dependent data resampling scheme, such as the moving block bootstrap. This, however, requires large sample sizes (say $n > 1000$). It is often worth the effort to attempt to linearise the problem and use the concept of residuals. Sometimes this approach is not possible in the time-domain. The alternative is to resample residuals, which are defined in the Fourier domain, and create new dependent observations from the resamples.

- We have discussed pivoting and variance stabilisation because they play a crucial role in ensuring statistical accuracy. This issue is further discussed in Chapter 3. It is worth noting that the variance stabilisation requires less computation than a nested bootstrap. This saving can be huge in some problems, such as confidence interval estimation, where one can reduce the number of resamples by a factor of ten.

Finally, we would like to direct the reader to other applications where the use of bootstrap techniques proved to be of significant value. It is worth mentioning the following references where the bootstrap has been used to improve the quality of parameter estimators (Arnholt *et al.*, 1998; Kim and Singh, 1998; Andrews, 2002) and to provide support in Bayesian analysis (Djurić, 1997; Linville *et al.*, 2001), especially for data with missing observations (Chung and Han, 2000; Hens *et al.*, 2001; Kim, 2002).

3
Signal detection with the bootstrap

Detection of signals in interference is a key area in signal processing applications such as radar, sonar and telecommunications. The theory of signal detection has been extensively covered in the literature. Many textbooks exist, including the classic by Van Trees (2001a) and his later additions to the series (Van Trees, 2001b, 2002a,b), the text on radar detection by DiFranco and Rubin (1980), and several texts on estimation and detection (Scharf, 1991; Poor, 1994; Kay, 1993, 1998). Signal detection theory is well established when the interference is Gaussian. However, methods for detection in the non-Gaussian case are often cumbersome and in many cases non-optimal.

Signal detection is formulated as a test of a hypothesis (Lehmann, 1991). To cover signal detection, we first need to introduce some concepts of hypothesis testing. This is followed by an exposition of bootstrap based hypothesis testing. In the second part of the chapter, we provide details on bootstrap detection of signals in Gaussian and non-Gaussian noise and show how bootstrap detection alleviates the restrictions imposed by classical detectors.

3.1 Principles of hypothesis testing

As the term suggests, in hypothesis testing one wishes to decide whether or not some formulated hypothesis is correct. The choice is between two decisions: accepting or rejecting the hypothesis. A decision procedure for such a problem is called a test of the hypothesis in question.

Let $f(\boldsymbol{x})$ be the probability density function of a random column vector $\boldsymbol{X} = (X_1, \ldots, X_n)'$. The density function can be considered as an element of a broader set of parameterised functions, defined as

$$\{f(\boldsymbol{x}|\boldsymbol{\theta}) : \boldsymbol{\theta} \in \Theta\},$$

where the parameter vector $\boldsymbol{\theta}$ of the parameter space Θ is unknown. The set of all possible values of $\boldsymbol{\theta}$ is divided into two mutually exclusive sets, Θ_H and Θ_K. These two sets are such that Θ_H is the set of all permitted values of $\boldsymbol{\theta}$ when the hypothesis is true, and Θ_K is the set of all permitted values of $\boldsymbol{\theta}$ in the alternative situation when the hypothesis is false. Thus, the hypothesis H is given by

$$\mathsf{H} : \boldsymbol{\theta} \in \Theta_H$$

and the alternative by

$$\mathsf{K} : \boldsymbol{\theta} \in \Theta_K \,.$$

In some cases, Θ_H is a set with only one element θ_0, and Θ_K contains all other permitted values, e.g.

$$\begin{aligned}\mathsf{H} : \quad & \theta = \theta_0, \\ \mathsf{K} : \quad & \theta > \theta_0.\end{aligned}$$

In this case, the hypothesis is referred to as a simple hypothesis while the alternative is composite.

Before we proceed with the actual principle of hypothesis testing, let us first consider a typical scenario of radar detection as shown in Figure 3.1.

Example 3.1.1 Radar detection

In radar we may wish to determine the presence or absence of an aircraft and subsequently to estimate its position, for example. We transmit an electromagnetic pulse that is reflected by the aircraft, causing an echo to be received by the antenna τ seconds later.

In the absence of background noise, detection poses no difficulty; that is, however small the reflected signal from a target, in theory it may be detected with sufficient gain in the receiver. Background noise, however, imposes a limit on the minimum detectable signal. The question of target existence is, in fact, a choice of deciding between noise alone or signal and noise. The noise background includes both random (thermal) noise, which is present in all systems, and often another type of noise, of an entirely different character, called clutter (Skolnik, 1990; Sekine and Mao, 1990). Random-noise interference exists in all physical devices at temperatures above absolute zero; clutter generally refers to unwanted back-scattered radar signals from ground objects, sea waves, rain, etc. Additionally, the signal may be corrupted by man-made interference (e.g. electromagnetic power line interference).

Once the presence of the object (target) has been established, its range,

Fig. 3.1. A typical scenario in radar detection.

r, is determined by the equation $\tau = 2r/c$, where c is the speed of electro-magnetic propagation. Clearly, if the round trip delay τ can be measured, then so can the range. The received echo is decreased in amplitude due to propagation losses and hence may be obscured by environmental noise. Its onset may also be perturbed by time delays introduced by the electronics of the receiver.

Suppose that the transmitted signal is of the form

$$s(t) = a(t) \cos \omega_0 t,$$

where $a(t)$ is the amplitude of the signal and ω_0 is the frequency of the radar electromagnetic pulse. The received signal within the time range $0 \le t \le T$ can be modelled by

$$X(t) = \begin{cases} \eta a(t - \tau) \cos[\omega_0(t - \tau) + \phi] + Z(t), & \text{target present} \\ \\ Z(t), & \text{otherwise.} \end{cases}$$

where $\eta \ge 0$ is an unknown amplitude, τ is the time delay and ϕ is an unknown phase. $Z(t)$ is noise, often assumed to be stationary. The problem in radar detection is to decide whether $\eta = 0$ or $\eta \neq 0$.

According to our previous notation, we can identify the following parameters and sets in this example:

$$
\begin{aligned}
\boldsymbol{\theta} &= (\eta, \tau, \phi)', \\
\Theta &= \{(\eta, \tau, \phi) : \eta \geq 0, \tau \geq 0, -\pi \leq \phi < \pi\}, \\
\Theta_H &= \{(0, \tau, \phi) : \tau \geq 0, -\pi \leq \phi < \pi\}, \\
\Theta_K &= \{(\eta, \tau, \phi) : \eta > 0, \tau \geq 0, -\pi \leq \phi < \pi\}.
\end{aligned}
$$

In practice some or all of the parameters may not be known *a priori* and need to be estimated before detection can be performed. We will return to this signal detection problem in Example 3.1.2. We now continue with hypothesis testing.

The problem is to decide whether to accept H (e.g. no signal present), i.e., $\boldsymbol{\theta} \in \Theta_H$, or reject H, i.e., $\boldsymbol{\theta} \in \Theta_K$ based on observations $\boldsymbol{x}$ of $\boldsymbol{X}$, where $\boldsymbol{x}$ is a collection of sampled values of the received continuous-time signal $x(t)$ and accordingly $\boldsymbol{X}$, its random counterpart, is a model for the observations. To intelligently make this decision, we will use the framework of statistical hypothesis testing.

To test the hypothesis H against the alternative K we define a function of the observations that is bounded, the so-called test function:

$$0 \leq \varepsilon(\boldsymbol{x}) \leq 1.$$

The function $\varepsilon(\boldsymbol{x})$ can be interpreted as a probability of deciding whether the parameter vector $\boldsymbol{\theta}$ lies in the parameter space of values permitted under the alternative K, i.e.,

$$\boldsymbol{\theta} \in \Theta_K,$$

in which case we say that we accept H with probability $1 - \varepsilon(\boldsymbol{x})$ and K with probability $\varepsilon(\boldsymbol{x})$. The test is then called randomised.

In some cases, the test function $\varepsilon(\boldsymbol{x})$ may accept only the values zero and one, and we then say that the test is non-randomised, with a critical region defined as

$$\mathcal{C} = \{\boldsymbol{x} : \varepsilon(\boldsymbol{x}) = 1\} = \{\boldsymbol{x} : T_n(\boldsymbol{x}) \in \mathcal{R}\}$$

where $\mathcal{R}$ is the rejection region and $T_n(\boldsymbol{x})$ is some other function of the observations. Its random counterpart $T_n(\boldsymbol{X})$ is referred to as the *test statistic*.

Denoting the acceptance region by $\mathcal{A}$ we have

$$\varepsilon(\boldsymbol{x}) = \left\{ \begin{array}{ll} 1, & \text{if } T_n(\boldsymbol{x}) \in \mathcal{R} \\ 0, & \text{if } T_n(\boldsymbol{x}) \in \mathcal{A} \end{array} \right. .$$

Such a deterministic test is referred to as a binary test.

The test function $\varepsilon(\boldsymbol{x})$ may not be unique and there might be several such functions used in a particular hypothesis testing problem. Thus, it is important to statistically evaluate the performance of the test. To do so, we calculate the so-called *power of the test*, defined by

$$\beta_\varepsilon(\boldsymbol{\theta}) = \mathsf{E}_{\boldsymbol{\theta}}[\varepsilon(\boldsymbol{X})] = \int_{-\infty}^{\infty} \varepsilon(\boldsymbol{x}) \cdot f(\boldsymbol{x}|\boldsymbol{\theta}) d\boldsymbol{x} \,,$$

which is essentially the probability that, independently of the observations $\boldsymbol{x}$, the test function $\varepsilon(\boldsymbol{x})$ decides for the alternative K when $\boldsymbol{\theta} \in \Theta_{\mathsf{K}}$. Herein $\mathsf{E}_{\boldsymbol{\theta}}$ is the expectation with respect to $f(\boldsymbol{x}|\boldsymbol{\theta})$.

Another important characteristic of a hypothesis test is the *level of significance*. The test $\varepsilon(\boldsymbol{x})$ is said to be a test of level or size α for $(\alpha, \mathsf{H}, \mathsf{K})$, if $\beta_\varepsilon(\boldsymbol{\theta}) \leq \alpha$ for all $\boldsymbol{\theta} \in \Theta_{\mathsf{H}}$.

In most applications, we wish to use a test that has the highest power. A test $\varepsilon_0(\boldsymbol{x})$ is called uniformly most powerful (UMP) for $(\alpha, \mathsf{H}, \mathsf{K})$ if the inequality

$$\beta_{\varepsilon_*}(\boldsymbol{\theta}) \geq \beta_\varepsilon(\boldsymbol{\theta}) \quad \text{for all} \quad \boldsymbol{\theta} \in \Theta_{\mathsf{K}}$$

holds given the tests are of level α. It should be noted that in many practical cases the most powerful test depends on the values of the parameter vector $\boldsymbol{\theta} \in \Theta_{\mathsf{K}}$ and according to the statement above it is not a UMP test. However, a UMP test exists if the distribution of the observations has a so-called monotone likelihood ratio (Lehmann, 1991; Scharf, 1991, p. 78 and p. 124, respectively). For example, the one-parameter exponential family of distributions with probability density function

$$f(\boldsymbol{x}|\theta) = a(\theta)b(\boldsymbol{x}) \exp[c(\theta)d(\boldsymbol{x})]$$

has a monotone likelihood ratio if $c(\theta)$ is a nondecreasing function (Scharf, 1991, p. 126).

As mentioned earlier, if Θ_{H} consists only of one element $\{\theta_0\}$ then H is called a simple hypothesis, otherwise it is called a composite hypothesis. A similar definition holds for K. For convenience let us denote $\Theta_{\mathsf{H}} = \{\theta_0\}$ and $\Theta_{\mathsf{K}} = \{\theta_1\}$. In this case we have to decide whether $f(\boldsymbol{x}) = f(\boldsymbol{x}|\theta_0)$ or $f(\boldsymbol{x}|\theta_1)$, where $f(\boldsymbol{x}|\theta_0)$ is the probability density function under the hypothesis while $f(\boldsymbol{x}|\theta_1)$ is the probability density function under the alternative.

In Example 3.1.1, η would take only the values $\eta = 0$ or $\eta = \eta_1$ if H and K were simple, which means that the signal to be detected is known.

In order to derive the most powerful test, we need to use an optimality criterion. Optimal hypothesis testing can be achieved using Bayes' or Neyman–Pearson's criterion (Neyman and Pearson, 1928). The Neyman–Pearson lemma tells us how to find the most powerful test of size α for testing the simple hypothesis H against the simple alternative K.

Theorem 3.1.1 *Let* $\Theta = \{\theta_0, \theta_1\}$ *and assume that the distribution of* $\boldsymbol{X}$ *has a probability density* $f(\boldsymbol{x}|\theta_i), i = 0, 1$. *Then,*

i) *There exists a test* $\varepsilon_0(\boldsymbol{x})$ *for* $(\alpha, \mathsf{H}, \mathsf{K})$ *and a constant* T_α *such that*

1. $\beta_{\varepsilon.}(\theta_0) = \displaystyle\int_{-\infty}^{\infty} \varepsilon_0(\boldsymbol{x})f(\boldsymbol{x}|\theta_0)d\boldsymbol{x} = \alpha$

2. $\varepsilon_0(\boldsymbol{x}) = \begin{cases} 1, & \text{when } f(\boldsymbol{x}|\theta_1) > T_\alpha f(\boldsymbol{x}|\theta_0) \\ 0, & \text{when } f(\boldsymbol{x}|\theta_1) < T_\alpha f(\boldsymbol{x}|\theta_0) \end{cases}$

ii) *If a test satisfies 1. and 2. for some* T_α, *then it is most powerful for* $(\alpha, \mathsf{H}, \mathsf{K})$.

iii) *If* $\varepsilon_0(\boldsymbol{x})$ *is most powerful for* $(\alpha, \mathsf{H}, \mathsf{K})$, *then for some* T_α *it satisfies 2. almost everywhere. It also satisfies 1. unless there exists a test of size less than* α *and with power 1.*

The proof of the lemma can be found in the text by Lehmann (1991).

Going back to our radar example, the hypothesis H is the situation where no signal is present while under the alternative K both signal and noise are present. The test function $\varepsilon(\boldsymbol{x})$ is what is called in the engineering (radar) community the detector. Often, we measure the probability of detection (P_D) and the probability of false alarm (P_F). These refer respectively to $\beta_\varepsilon(\boldsymbol{\theta})$, when $\boldsymbol{\theta} \in \Theta_\mathsf{K}$ and $\beta_\varepsilon(\boldsymbol{\theta})$, when $\boldsymbol{\theta} \in \Theta_\mathsf{H}$.

Using this terminology, we can interpret the Neyman–Pearson criterion as the foundation for an optimal test that maximises the probability of detection while maintaining the level of false alarm to a preset value.

The problem of signal detection can now be posed in the following framework. Let

$$\begin{aligned} \mathsf{H} &: \quad \boldsymbol{\theta} \in \Theta_\mathsf{H} = \{\boldsymbol{\theta}_0\} \\ \mathsf{K} &: \quad \boldsymbol{\theta} \in \Theta_\mathsf{K} = \{\boldsymbol{\theta}_1\} \end{aligned}$$

We wish to design a detector $\varepsilon(\boldsymbol{x})$ for $(\alpha, \mathsf{H}, \mathsf{K})$ where the probability of false alarm $\mathsf{P}_F = \beta_\varepsilon(\boldsymbol{\theta}_0) = \alpha$ is given. We are interested in the probability of detection $\mathsf{P}_D = \beta_\varepsilon(\boldsymbol{\theta}_1)$. Note that the probability of detection is a function of α, $\mathsf{P}_D = \mathsf{P}_D(\alpha)$, and is called a receiver operating characteristic (ROC). If $\varepsilon(\boldsymbol{x})$ is UMP for $(\alpha, \mathsf{H}, \mathsf{K})$ the following properties hold for $\mathsf{P}_D(\alpha)$:

(i) $\mathsf{P}_D(\alpha)$ is nondecreasing and concave in $0 \le \alpha \le 1$

(ii) $\mathsf{P}_D(\alpha)$ is continuous in $0 \le \alpha \le 1$

(iii) $\dfrac{\mathsf{P}_D(\alpha)}{\alpha}$ is nonincreasing in $0 < \alpha < 1$

(iv) $\lim\limits_{\alpha \to 1} \dfrac{\mathsf{P}_D(\alpha)}{\alpha} = 1$

(v) $1 \le \dfrac{\mathsf{P}_D(\alpha)}{\alpha} \le 1/\alpha$

Let us consider a typical signal detection problem as illustrated in Figure 3.2.

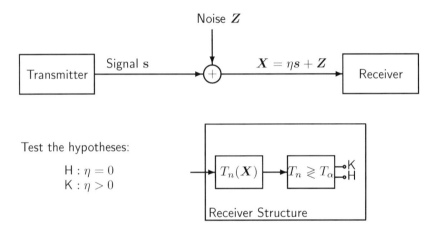

Fig. 3.2. A typical signal detection problem.

Example 3.1.2 A simple signal detection problem

Following the notation from Figure 3.2, consider the model

$$X_t = \eta \cdot s_t + Z_t, \qquad t = 1, \ldots, n$$

or alternatively using vector notation

$$\boldsymbol{X} = \eta \cdot \boldsymbol{s} + \boldsymbol{Z},$$

where $\boldsymbol{X} = (X_1, \ldots, X_n)'$ is a collection of independent random variables. Herein s_t is a known signal, e.g. (see Example 3.1.1)

$$s_t = a_t \cos(\omega_0 t),$$

where a_t is a sampled version of $a(t)$, $\eta \ge 0$ is the unknown signal amplitude and Z_t is white noise with $\mathsf{E}[Z_t] = 0$ and a known variance $\mathsf{var}[Z_t] = \sigma_Z^2$. Assume that the noise is normally distributed, i.e. $\boldsymbol{Z} \sim \mathcal{N}(\boldsymbol{0}, \sigma_Z^2 \boldsymbol{I})$, where $\boldsymbol{0}$ is the zero vector and $\boldsymbol{I}$ is the identity matrix.

The signal detection problem can be formulated in the following way. Given observations $x_1, \ldots, x_n$ of $X_1, \ldots, X_n$, decide whether $\boldsymbol{X} = \eta \cdot \boldsymbol{s} + \boldsymbol{Z}$ or $\boldsymbol{X} = \boldsymbol{Z}$, i.e. whether $\eta > 0$ or $\eta = 0$. We perform the following steps.

(i) Define the hypothesis H:

The random vector $\boldsymbol{X}$ does not contain a signal $\boldsymbol{s}$. Thus, $\boldsymbol{X} = \boldsymbol{Z}$ or equivalently $\eta = 0$. In this case the observations are normally distributed with zero-mean and variance σ_Z^2, i.e. $\boldsymbol{X} \sim \mathcal{N}(\boldsymbol{0}, \sigma_Z^2 \boldsymbol{I})$.

(ii) Define the alternative K:

The random vector $\boldsymbol{X}$ contains the signal $\boldsymbol{s}$, i.e. $\boldsymbol{X} = \eta \boldsymbol{s} + \boldsymbol{Z}$, where η is unknown but $\eta > 0$. Then, $\boldsymbol{X} \sim \mathcal{N}(\eta \boldsymbol{s}, \sigma_Z^2 \boldsymbol{I})$.

(iii) Choose an appropriate function $T_n(\boldsymbol{x})$ of the observations $\boldsymbol{x} = (x_1, \ldots, x_n)'$ to test H:

This can be

$$T_n(\boldsymbol{x}) = \boldsymbol{s}' \boldsymbol{x}.$$

(iv) Determine the distribution of the test statistic $T_n = T_n(\boldsymbol{X})$ under H:

It is straightforward to show that under H, $T_n = T_n(\boldsymbol{X}) = \boldsymbol{s}' \boldsymbol{X}$ is normally distributed with

$$\mathsf{E}[T_n] = \mathsf{E}[\boldsymbol{s}' \boldsymbol{X}] = \boldsymbol{s}' \mathsf{E}[\boldsymbol{X}] = 0$$

and

$$\begin{aligned}
\mathsf{var}[T_n] &= \sigma_T^2 \\
&= \mathsf{E}[T_n - \mathsf{E}[T_n]]^2 = \mathsf{E}[T_n^2] \\
&= \mathsf{E}[\boldsymbol{s}' \boldsymbol{X} \boldsymbol{X}' \boldsymbol{s}] = \boldsymbol{s}' \mathsf{E}[\boldsymbol{X} \boldsymbol{X}'] \boldsymbol{s} \\
&= \boldsymbol{s}'(\sigma_Z^2 \boldsymbol{I}) \boldsymbol{s} = \sigma_Z^2 \boldsymbol{s}' \boldsymbol{s}.
\end{aligned}$$

(v) Determine the critical region $\mathcal{C} \in \mathbb{R}^n$ of observation $\boldsymbol{x}$, for which H is rejected:

Note first that if the number $0 < \alpha < 1$ is a predetermined level of significance then $\mathcal{C}$ is determined so that

$$\alpha = \mathsf{Prob}[\boldsymbol{X} \in \mathcal{C}|\mathsf{H}] = \mathsf{Prob}[T_n \in \mathcal{R}|\mathsf{H}]$$

and $\mathcal{C} = \{x : T_n(x) \in \mathcal{R}\}$. We then have

$$\alpha = \mathsf{Prob}[T_n > T_\alpha | \mathsf{H}] = \mathsf{Prob}\left[\frac{T_n}{\sigma_T} > \frac{T_\alpha}{\sigma_T} | \mathsf{H}\right] = 1 - \mathsf{Prob}\left[U \leq \frac{T_\alpha}{\sigma_T}\right],$$

where $U = \frac{T_n}{\sigma_T} \sim \mathcal{N}(0,1)$. The parameter T_α is obtained so that

$$\Phi\left(\frac{T_\alpha}{\sigma_T}\right) = 1 - \alpha,$$

where Φ is the cumulative distribution function of the standard Gaussian distribution.

(vi) Given $x_1, \ldots, x_n$ and α the test consists of either rejecting H if $T_n(x) \in \mathcal{R}$ or accepting it if $T_n(x) \notin \mathcal{R}$.

Given α, we decide K, i.e. X contains a signal s (or we detect a signal s) if $T_n(x) > T_\alpha$. Alternatively, we decide H, i.e. X contains no signal s if $T_n(x) \leq T_\alpha$.

The above example shows the single steps required in a detection problem. We would like to stress that in order to solve this relatively simple problem we need to know the noise distribution. In many practical cases the interference can simply not be modelled by a Gaussian distribution and other techniques have to be used instead. However, before we proceed any further, let us turn our attention to a performance analysis of the above detector.

Calculation of T_α and $\mathsf{P}_D = \mathsf{P}_D(\alpha)$. We showed that

$$\alpha = 1 - \Phi(T_\alpha/\sigma_T) = 1 - \Phi(N_\alpha)$$

where $N_\alpha = T_\alpha/\sigma_T$ is the value that with probability α is exceeded by a standard Gaussian distributed random variable. Then, $T_\alpha = \exp\{N_\alpha\sqrt{s's} - s's/2\}$. We define an appropriate parameter $d^2 = s's = s's/\sigma_Z^2$, which is proportional to the signal-to-noise power ratio (SNR), namely $n \cdot$SNR. Then we can express T_α as

$$T_\alpha = \exp\{N_\alpha d - d^2/2\}.$$

The probability of detection can be obtained by

$$\begin{aligned}\mathsf{P}_D &= \mathsf{Pr}\{T_n > T_\alpha | \mathsf{K}\} = 1 - \Phi(T_\alpha - s's/\sigma_T) \\ &= 1 - \Phi(N_\alpha - d)\end{aligned}$$

A plot of P_D as a function of α is given in Figure 3.3.

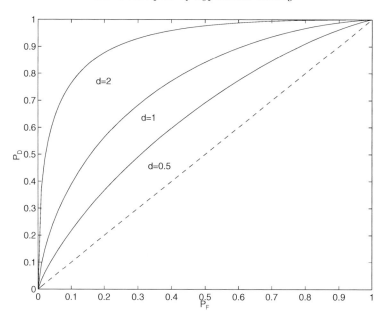

Fig. 3.3. ROC of the detector in Example 3.1.2.

Example 3.1.3 The likelihood ratio

Consider the test which decides the hypothesis $\boldsymbol{\theta} = \boldsymbol{\theta}_0$ against the alternative $\boldsymbol{\theta} = \boldsymbol{\theta}_1$, given Gaussian data $\boldsymbol{X}$. Then, $f(\boldsymbol{x}|\boldsymbol{\theta}_0)$ is the density of $\mathcal{N}(\boldsymbol{\mu}_0, \boldsymbol{C}_0)$ and $f(\boldsymbol{x}|\boldsymbol{\theta}_1)$ is the density of $\mathcal{N}(\boldsymbol{\mu}_1, \boldsymbol{C}_1)$. Following Neyman–Pearson's lemma, a UMP test checks whether $f(\boldsymbol{x}|\boldsymbol{\theta}_1) \gtrless f(\boldsymbol{x}|\boldsymbol{\theta}_0)$. Because for the Gaussian distribution $f_{\boldsymbol{X}}(\boldsymbol{x}) > 0$, the test checks whether

$$\Lambda(\boldsymbol{x}) = \frac{f(\boldsymbol{x}|\boldsymbol{\theta}_1)}{f(\boldsymbol{x}|\boldsymbol{\theta}_0)} \gtrless T_\alpha$$

or equivalently

$$\ln \Lambda(\boldsymbol{x}) = \ln f(\boldsymbol{x}|\boldsymbol{\theta}_1) - \ln f(\boldsymbol{x}|\boldsymbol{\theta}_0) \gtrless \ln T_\alpha \,,$$

where $\Lambda(\boldsymbol{x})$ is called the likelihood ratio and $\ln \Lambda(\boldsymbol{x})$ is the log-likelihood ratio. A likelihood ratio can also be formulated for complex random vectors. When the random vector is circularly symmetric, the likelihood ratio is often formulated in terms of the first order amplitude probability density functions (Conte *et al.*, 1995).

For Gaussian densities, the log-likelihood ratio leads to

$$\ln \Lambda(\boldsymbol{x}) = \frac{1}{2}(\ln |\boldsymbol{C}_0| - \ln |\boldsymbol{C}_1| + \boldsymbol{\mu}_0' \boldsymbol{C}_0^{-1} \boldsymbol{\mu}_0 - \boldsymbol{\mu}_1' \boldsymbol{C}_1^{-1} \boldsymbol{\mu}_1)$$

$$+ (\boldsymbol{\mu}_1' \boldsymbol{C}_1^{-1} - \boldsymbol{\mu}_0' \boldsymbol{C}_0^{-1})\boldsymbol{x} + \frac{1}{2}\boldsymbol{x}'(\boldsymbol{C}_0^{-1} - \boldsymbol{C}_1^{-1})\boldsymbol{x}$$

A uniformly most powerful test for $(\alpha, \mathsf{H}, \mathsf{K})$ is then given by

$$\varepsilon_0(\boldsymbol{x}) = \begin{cases} 1, & \text{when } \ln \Lambda(\boldsymbol{x}) > \ln T_\alpha \\ & \text{or } T_n(\boldsymbol{x}) = (\boldsymbol{\mu}_1' \boldsymbol{C}_1^{-1} - \boldsymbol{\mu}_0' \boldsymbol{C}_0^{-1})\boldsymbol{x} + \frac{1}{2}\boldsymbol{x}'(\boldsymbol{C}_0^{-1} - \boldsymbol{C}_1^{-1})\boldsymbol{x} > T_\alpha' \\ 0, & \text{otherwise} \end{cases}$$

for a suitable $T_\alpha = T_\alpha(\alpha)$, provided $\boldsymbol{\mu}_0 \neq \boldsymbol{\mu}_1$, and $\boldsymbol{C}_0 \neq \boldsymbol{C}_1$ with $T_\alpha' = \ln T_\alpha - \frac{1}{2}(\ln |\boldsymbol{C}_0| - \ln |\boldsymbol{C}_1| + \boldsymbol{\mu}_0' \boldsymbol{C}_0^{-1} \boldsymbol{\mu}_0 - \boldsymbol{\mu}_1' \boldsymbol{C}_1^{-1} \boldsymbol{\mu}_1)$.

3.1.1 Sub-optimal detection

In the Neyman–Pearson criterion, the distribution under the hypothesis is assumed to be known. Note however that in many practical signal detection problems it is not possible to derive parametrically or otherwise a UMP test, especially in the case where the interference is non-Gaussian. For example, if $\boldsymbol{\theta}$ is not random and the likelihood ratio is different for a different $\boldsymbol{\theta}_1$, then the test statistic is not fully specified and a detector that satisfies the Neyman Pearson criterion cannot be constructed. When the knowledge of the distribution under the hypothesis is not available, we resort to sub-optimal methods based on non-parametric or robust statistics (Thomas, 1970; Miller and Thomas, 1972; Poor, 1986; Kazakos and Papantoni-Kazakos, 1990; Gibson and Melsa, 1996).

The most often used sub-optimal method is the generalised likelihood ratio test with the test statistic:

$$\Lambda(\boldsymbol{x}) = \frac{f(\boldsymbol{x}|\hat{\boldsymbol{\theta}}_1)}{f(\boldsymbol{x}|\boldsymbol{\theta}_0)} \gtrless T_\alpha \tag{3.1}$$

where $\hat{\boldsymbol{\theta}}_1$ is the maximum likelihood estimator for $\boldsymbol{\theta}_1$. One property of the generalised likelihood ratio test is that, under some regularity conditions, it is consistent where consistency is defined as follows.

Definition 3.1.1 (Consistency) *A detector is consistent if*

$$\lim_{n\to\infty} \mathsf{P}_D \to 1. \tag{3.2}$$

If the distribution of the interference deviates from the assumed distribution, then it is also meaningful to consider the *robustness* of the test.

Definition 3.1.2 (Robustness) *A detector is robust if (Gibson and Melsa, 1996)*

$$\lim_{n \to \infty} \mathsf{P}_F \to \alpha. \tag{3.3}$$

The detector is also known as asymptotically nonparametric.

These two properties are desirable when an optimal detector that satisfies the Neyman–Pearson criterion does not exist. The extent to which these properties apply to bootstrap methods for signal detection is examined later in this chapter.

3.2 Hypothesis testing with the bootstrap

Two major problems can be encountered in hypothesis testing. The first one occurs when the size of the random sample is small and asymptotic methods do not apply. The second possible problem is that the distribution of the test statistic cannot be determined analytically. One can overcome both problems by using bootstrap techniques.

Consider a situation in which a random sample $\mathcal{X} = \{X_1, \ldots, X_n\}$ is observed from its unspecified probability distribution F_θ, where θ, a characteristic of F, is unknown. We are interested in testing the hypothesis

$$\mathsf{H} : \theta \le \theta_0 \qquad \text{against the alternative} \qquad \mathsf{K} : \theta > \theta_0 \,,$$

where θ_0 is some known bound. Let $\hat{\theta}$ be an estimator of θ and $\hat{\sigma}^2$ an estimator of the variance σ^2 of $\hat{\theta}$.

Define the test statistic

$$T_n = \frac{\hat{\theta} - \theta_0}{\hat{\sigma}}. \tag{3.4}$$

The inclusion of the scale factor $\hat{\sigma}$, to be defined in Section 3.4, ensures, as discussed in Section 2.2, that T_n is asymptotically pivotal as $n \to \infty$ (Hall and Titterington, 1989; Hall and Wilson, 1991). Using a pivotal statistic means we only need to deal with the appropriate standard distribution rather than a whole family of distributions. Although not recommended, the scale factor $\hat{\sigma}$ can be omitted in some applications. This often brings a significant computational saving. We will address this problem in Example 3.4.2.

If the distribution function G of T_n under H were known then an exact α-level test would suggest rejecting H if $T_n \ge T_\alpha$, where, as discussed earlier,

T_α is determined by $G(T_\alpha) = 1 - \alpha$ (Lehmann, 1991). For example, if F has mean μ and unknown variance,

$$
T_n = \frac{\dfrac{1}{n} \displaystyle\sum_{i=1}^{n} X_i - \mu_0}{\sqrt{\dfrac{1}{n(n-1)} \displaystyle\sum_{i=1}^{n} \left(X_i - \dfrac{1}{n} \displaystyle\sum_{j=1}^{n} X_j \right)^2}}
$$

is used to test $\mu \leq \mu_0$ against $\mu > \mu_0$, given the random sample $\mathcal{X} = \{X_1, X_2, \ldots, X_n\}$. For large n, T_n is asymptotically t-distributed with $n-1$ degrees of freedom.

The bootstrap approach for testing $\mathsf{H} : \theta \leq \theta_0$ against $\mathsf{K} : \theta > \theta_0$, given $\hat{\theta}$ and $\hat{\sigma}$ (see Section 3.4), found from $\mathcal{X}$, is illustrated in Figure 3.4 and Table 3.1. In the approach, we retain the asymptotically pivotal nature of the test statistic because the bootstrap approximation of the distribution of T_n is better than the approximation of the distribution of $\hat{\theta}$ (Hall, 1988). Note that in Step 2 of Table 3.1 the constant θ_0 has been replaced in (3.5) by the estimate of θ, $\hat{\theta}$, derived from $\mathcal{X}$. This is crucial if the test is to have good power properties. It is also important in the context of the accuracy of the level of the test (Hall and Titterington, 1989; Hall and Wilson, 1991; Hall, 1992; Zoubir, 1994).

Note that in the case where one is interested in the hypothesis $\mathsf{H} : \theta = \theta_0$ against the two-sided alternative $\mathsf{K} : \theta \neq \theta_0$, the procedure shown in Table 3.1 is still valid, except that $\hat{\theta} - \theta_0$ is replaced by $|\hat{\theta} - \theta_0|$ in (3.4) so that T_n is given by $T_n = |\hat{\theta} - \theta_0|/\hat{\sigma}$ and correspondingly $T_n^* = |\hat{\theta}^* - \hat{\theta}|/\hat{\sigma}^*$ (Hall and Titterington, 1989).

3.3 The role of pivoting

We mentioned earlier (see Section 2.2) the importance of pivoting in the context of confidence interval estimation. Let us re-emphasise this concept in a more formal manner. Suppose that the distribution of the statistic T_n under H admits the Edgeworth expansion (Hall, 1988, 1992)

$$
G(x) = \mathsf{Prob}[T_n \leq x] = \Phi(x) + n^{-1/2}q(x)\phi(x) + O(n^{-1}), \tag{3.6}
$$

where q is an even quadratic polynomial while Φ and ϕ are the standard Gaussian distribution and density functions, respectively. The bootstrap estimate of G admits an analogous expansion,

$$
\hat{G}(x) = \mathsf{Prob}[T_n^* \leq x | \mathcal{X}] = \Phi(x) + n^{-1/2}\hat{q}(x)\phi(x) + O_p(n^{-1}),
$$

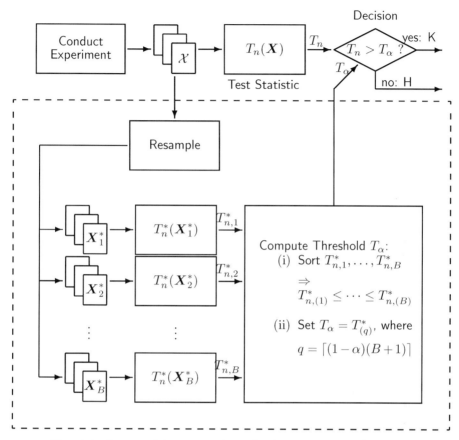

Fig. 3.4. A general bootstrap hypothesis testing procedure.

where T_n^* is the bootstrap version of T_n and the polynomial $\hat{q}$ is obtained from q by replacing unknowns, such as skewness, by bootstrap estimates. The term $O_p(n^{-1})$ denotes a random variable that is of order n^{-1} "in probability". The distribution of T_n^* conditional on $\mathcal{X}$ is called the bootstrap distribution of T_n. It is known that $\hat{q} - q = O_p(n^{-1/2})$ (Hall, 1988) and thus,

$$\mathsf{Prob}[T_n^* \leq x | \mathcal{X}] - \mathsf{Prob}[T_n \leq x] = O_p(n^{-1}).$$

That is, the bootstrap approximation to G is in error by n^{-1}. If we had approximated G by Φ, the error would have been of order $n^{-1/2}$ as it can be seen from (3.6). Thus, the difference between the actual level and the nominal level of the test is $\alpha' - \alpha = O(n^{-1})$ (Hall and Titterington, 1989;

Table 3.1. *The bootstrap principle for testing the hypothesis* $\mathsf{H} : \theta \leq \theta_0$
against $\mathsf{K} : \theta > \theta_0$.

Step 0. *Experiment.* Conduct the experiment and collect the data into the sample $\mathcal{X} = \{X_1, \ldots, X_n\}$.

Step 1. *Resampling.* Draw a random sample $\mathcal{X}^*$ of the same size as $\mathcal{X}$, with replacement, from $\mathcal{X}$.

Step 2. *Calculation of the bootstrap statistic.* From $\mathcal{X}^*$, calculate

$$T_n^* = \frac{\hat{\theta}^* - \hat{\theta}}{\hat{\sigma}^*}, \qquad (3.5)$$

where $\hat{\theta}$ replaces θ_0, and $\hat{\theta}^*$ and $\hat{\sigma}^*$ are versions of $\hat{\theta}$ and $\hat{\sigma}$ computed in the same manner as $\hat{\theta}$ and $\hat{\sigma}$, respectively, but with the resample $\mathcal{X}^*$ replacing the sample $\mathcal{X}$.

Step 3. *Repetition.* Repeat Steps 1 and 2 many times to obtain a total of B bootstrap statistics $T_{n,1}^*, T_{n,2}^*, \ldots, T_{n,B}^*$.

Step 4. *Ranking.* Rank the collection $T_{n,1}^*, T_{n,2}^*, \ldots, T_{n,B}^*$ into increasing order to obtain $T_{n,(1)}^* \leq T_{n,(2)}^* \leq \cdots \leq T_{n,(B)}^*$.

Step 5. *Test.* A bootstrap test has then the following form: reject H if $T_n > T_{(q)}^*$, where the choice of q determines the level of significance of the test and is given by $\alpha = (B + 1 - q)(B + 1)^{-1}$, where α is the nominal level of significance (Hall and Titterington, 1989).

Hall, 1988, 1992). This result holds whenever the statistic is asymptotically pivotal.

To appreciate why pivoting yields a smaller error, let $U_n = \sqrt{n}(\hat{\theta} - \theta_0)$ be a non-pivotal statistic whose distribution we want to approximate by the bootstrap. In this case, we have (Hall, 1992)

$$
\begin{aligned}
F(x) &= \mathsf{Prob}[U_n \leq x] \\
&= \Phi(x/\sigma) + n^{-1/2} p(x/\sigma) \phi(x/\sigma) + O(n^{-1})
\end{aligned}
$$

and

$$
\begin{aligned}
\hat{F}(x) &= \mathsf{Prob}[U_n^* \leq x | \mathcal{X}] \\
&= \Phi(x/\hat{\sigma}) + n^{-1/2} \hat{p}(x/\sigma) \phi(x/\hat{\sigma}) + O_p(n^{-1}),
\end{aligned}
$$

where p is a polynomial, $\hat{p}$ is obtained from p by replacing unknowns by their bootstrap estimates, σ^2 equals the asymptotic variance of U_n, $\hat{\sigma}^2$ is the bootstrap estimator of σ^2, and U_n^* is the bootstrap version of U_n. Again, $\hat{p} - p = O_p(n^{-1/2})$, and also $\hat{\sigma} - \sigma = O_p(n^{-1/2})$. Thus, we have

$$\hat{F}(x) - F(x) = \Phi(x/\hat{\sigma}) - \Phi(x/\sigma) + O_p(n^{-1}).$$

The difference between $\hat{\sigma}$ and σ is usually of order $n^{-1/2}$. Indeed, $n^{-1/2}(\hat{\sigma} - \sigma)$ typically has a limiting Gaussian distribution $\mathcal{N}(0, \xi^2)$, for some $\xi > 0$. Thus, $\Phi(x/\hat{\sigma}) - \Phi(x/\sigma)$ is of size $n^{-1/2}$, not n^{-1}, and therefore the bootstrap approximation to F is in error by terms of size $n^{-1/2}$ (Hall, 1992).

Note that in practice it may not be easy to show for a given estimator $\hat{\theta}$ that the percentile-t or any other pivot is in fact pivotal. If the estimator can be put in the form of a smooth function of means or other statistics for which convergence results are available, the pivot method can be applied. If not, caution should be exercised. If a chosen pivot is in fact not pivotal, the pivot method will not work. This echoes the view of Schenker (1985) for bootstrap confidence intervals.

Example 3.3.1 Limitations of pivoting

Consider the following standardised pivot

$$\frac{\hat{\theta} - \theta}{\sigma_{\hat{\theta}}}, \tag{3.7}$$

and the test statistic, $T_n = (\hat{\theta} - \theta_0)/\sigma_{\hat{\theta}}$. Suppose the signal-plus-interference model is

$$X_t = \pm\sqrt{\theta} + Z_t, \qquad t = 1, \ldots, n, \tag{3.8}$$

where $\pm\sqrt{\theta}$ is the signal, which is independent of t, and $Z_1, \ldots, Z_n$ is iid, zero-mean interference with a variance σ^2. This model is considered in an estimation problem by Kay (1993, p. 174). We wish to test $\mathsf{H} : \theta = 0$ against $\mathsf{K} : \theta > 0$ for detection. In this case, $\theta_0 = 0$.

The maximum likelihood estimator $\hat{\theta} = n^{-1} \sum_t X^2$ has, asymptotically, the distribution of a $\frac{\sigma}{n}\chi_1^2(\theta/2)$-variate where $\chi_1^2(\theta/2)$ is a chi-square random variable with one degree of freedom and noncentrality parameter $\theta/2$. In addition, $\sigma_{\hat{\theta}}^2 = 2\frac{\sigma}{n}\left(\frac{\sigma}{n} + 2\theta\right)$ (Johnson and Kotz, 1970). The expectation $\mathsf{E}[\hat{\theta}]$ is $\frac{\sigma}{n} + \theta$. Therefore, the pivot is

$$\frac{\frac{\sigma}{n}\chi_1^2(\theta/2) - \theta}{\sqrt{2\frac{\sigma}{n}\left(\frac{\sigma}{n} + 2\theta\right)}} = \frac{\chi_1^2(\theta/2) - n\theta/\sigma^2}{\sqrt{2\left(1 + 2n\theta/\sigma^2\right)}}. \tag{3.9}$$

As $n \to \infty$, this becomes a standard Gaussian random variable for $\theta > 0$ and a $\frac{1}{\sqrt{2}}\chi_1^2$ random variable for $\theta = 0$. Therefore, since the distributions are different for different θ, (3.7) is not pivotal.

3.4 Variance estimation

The tests described in Table 3.1 of Section 3.2 require the estimation of $\hat{\sigma}$ and its bootstrap counterpart $\hat{\sigma}^*$. In this section, we discuss how one can estimate the parameters using the bootstrap.

Suppose that X is a real-valued random variable with unknown probability distribution F with mean μ_X and variance σ_X^2. Let $\mathcal{X} = \{X_1, X_2, \ldots, X_n\}$ be a random sample of size n from F. We wish to estimate μ_X and assign to it a measure of accuracy.

The sample mean $\hat{\mu}_X = n^{-1}\sum_{i=1}^n X_i$ is a natural estimate for μ_X which has expectation μ_X and variance σ_X^2/n. The standard deviation of the sample mean $\hat{\mu}_X$ is the square root of its variance $\sigma_X/\sqrt{n}$ and is the most common way of indicating statistical accuracy. The mean value and standard deviation of $\hat{\mu}_X$ are exact but the usual assumption of Gaussianity of $\hat{\mu}_X$ is an approximation only and is valid under general conditions on F as n grows.

In this example, we could use

$$\hat{\sigma}_X = \sqrt{\frac{1}{n-1}\sum_{i=1}^n (X_i - \hat{\mu}_X)^2} \qquad (3.10)$$

to estimate $\sigma_X = \sqrt{\mathsf{E}[X - \mu_X]^2}$. This gives the following estimate of the standard deviation of $\hat{\mu}_X$,

$$\hat{\sigma} = \hat{\sigma}_X/\sqrt{n} = \sqrt{\frac{1}{n(n-1)}\sum_{i=1}^n (X_i - \hat{\mu}_X)^2}. \qquad (3.11)$$

We have considered a simple example where $\hat{\theta} = \hat{\mu}_X$. Herein, the estimate of the standard deviation is the usual estimate of the standard deviation of the distribution F. However, for a more complicated $\hat{\theta}$, other than $\hat{\mu}_X$, it will be difficult to find such a formula that would enable the calculation of the estimate exactly. The bootstrap can be used to estimate $\sigma_{\hat{\theta}}$, the standard deviation of $\hat{\theta}$, without a theoretical calculation, irrespective of the complicated nature of the estimate, such as a spectral density. The procedure to estimate $\hat{\sigma}$, the standard deviation of $\hat{\theta}$ is equivalent to the one given in Table 2.3 in Chapter 2 (Hall, 1992; Efron and Tibshirani, 1993; Politis, 1998).

For the estimation of $\hat{\sigma}^*$, the divisor in $T_n^* = (\hat{\theta}^* - \hat{\theta})/\hat{\sigma}_{\hat{\theta}}^*$, we would proceed similarly as in Table 2.3, except that the procedure involves two nested levels of resampling. Herein, for each resample $\mathcal{X}_b^*$, $b = 1, \ldots, B_1$ we draw resamples $\mathcal{X}_b^{**}$, $b = 1, \ldots, B_2$, evaluate $\hat{\theta}_b^{**}$ from each resample to obtain B_2 replications, and calculate

$$\hat{\sigma} = \sqrt{\frac{1}{B-1} \sum_{i=1}^{B} \left(\hat{\theta}_i^* - \frac{1}{B} \sum_{j=1}^{B} \hat{\theta}_j^* \right)^2}$$

replacing $\hat{\theta}_b^*$ and B by $\hat{\theta}_b^{**}$ and B_2, respectively.

There are situations in which estimating the variance with a nested bootstrap may not be feasible. In such cases methods are being developed to reduce the number of computations (Karlsson and Löthgren, 2000). The jackknife (Miller, 1974; Politis, 1998) is another technique for estimating the standard deviation. As an alternative to the bootstrap, the jackknife method can be thought of as drawing n samples of size $n - 1$ each *without* replacement from the original sample of size n.

Suppose we are given the sample $\mathcal{X} = \{X_1, \ldots, X_n\}$ and an estimate $\hat{\theta}$ from $\mathcal{X}$. The jackknife method is based on the sample *delete-one observation at a time*,

$$\mathcal{X}^{(i)} = \{X_1, X_2, \ldots, X_{i-1}, X_{i+1}, \ldots, X_n\}$$

for $i = 1, 2, \ldots, n$ and is called the jackknife sample (Efron and Tibshirani, 1993). The ith jackknife sample consists of the data set with the ith observation removed. For each ith jackknife sample, we calculate the ith jackknife estimate $\hat{\theta}^{(i)}$ of θ, $i = 1, \ldots, n$. The jackknife estimate of standard deviation of $\hat{\theta}$ is

$$\hat{\sigma} = \sqrt{\frac{n-1}{n} \sum_{i=1}^{n} \left(\hat{\theta}^{(i)} - n^{-1} \sum_{j=1}^{n} \hat{\theta}^{(j)} \right)^2}. \tag{3.12}$$

The jackknife is computationally less expensive if n is less than the number of replicates used by the bootstrap for standard deviation estimation because it requires computation of $\hat{\theta}$ only for the n jackknife data sets. For example, if $B = 25$ resamples are necessary for standard deviation estimation with the bootstrap, and the sample size is $n = 10$, then clearly the jackknife would be computationally less expensive than the bootstrap.

Example 3.4.1 Test of equality of means (Behrens–Fisher Problem)

Consider first a classical statistical hypothesis testing problem in which an explicit form for the distribution of the test statistic under the null hypothesis cannot be derived. This is one of the class of problems that can be easily resolved with the bootstrap. Given real-valued random samples $\mathcal{X} = \{X_1, \ldots, X_n\}$ and $\mathcal{Y} = \{Y_1, \ldots, Y_m\}$ from Gaussian distributions with respective means μ_X and μ_Y and variances σ_X^2 and σ_Y^2, one is interested in testing the hypothesis $\mathsf{H} : \mu_X = \mu_Y$ against the two-sided alternative $\mathsf{K} : \mu_X \neq \mu_Y$. A test with the statistic

$$T_n = \frac{(\hat{\mu}_X - \hat{\mu}_Y) - (\mu_X - \mu_Y)}{\sqrt{\dfrac{\hat{\sigma}_X^2}{n-1} + \dfrac{\hat{\sigma}_Y^2}{m-1}}} \tag{3.13}$$

can be used for this purpose (Kendall and Stuart, 1967; Lehmann, 1991). In (3.13), $\hat{\mu}_X = n^{-1} \sum_{i=1}^{n} X_i$, $\hat{\mu}_Y = m^{-1} \sum_{i=1}^{m} Y_i$ are the respective means of the random samples $\mathcal{X}$ and $\mathcal{Y}$ and $\hat{\sigma}_X^2 = n^{-1} \sum_{i=1}^{n} (X_i - \hat{\mu}_X)^2$ and $\hat{\sigma}_Y^2 = m^{-1} \sum_{i=1}^{m} (Y_i - \hat{\mu}_Y)^2$ are the sample variances.

If $\sigma_X^2 = \sigma_Y^2$, T_n admits, under H, a t-distribution with $m + n - 2$ degrees of freedom. Otherwise, T_n is a mixture of two t-distributions. Then, T_n could be expressed as $T_n = T_1 \sin U + T_2 \cos U$, where T_1, T_2 and U are independent random variables, T_1 and T_2 have t-distributions with $n - 1$ and $m - 1$ degrees of freedom, respectively, and U is uniformly distributed over $(-\pi, \pi]$. In this case an explicit solution for the distribution of T_n cannot be derived (Kendall and Stuart, 1967). This problem also occurs if one or both distributions of the populations are not Gaussian.

We now discuss how the bootstrap can be used to test for equality of means without *a priori* knowledge of the distribution of T_n. Let $\mathcal{X}^* = \{X_1^*, \ldots, X_n^*\}$ and $\mathcal{Y}^* = \{Y_1^*, \ldots, Y_m^*\}$ be resamples drawn independently and randomly with replacement from $\mathcal{X}$ and $\mathcal{Y}$, respectively. Let $\hat{\mu}_X^* = n^{-1} \sum_{i=1}^{n} X_i^*$, $\hat{\mu}_Y^* = m^{-1} \sum_{i=1}^{m} Y_i^*$, $\hat{\sigma}_X^{*2} = n^{-1} \sum_{i=1}^{n} (X_i^* - \hat{\mu}_X^*)^2$ and $\hat{\sigma}_Y^{*2} = m^{-1} \sum_{i=1}^{m} (Y_i^* - \hat{\mu}_Y^*)^2$ be the bootstrap analogues of $\hat{\mu}_X$, $\hat{\mu}_Y$, $\hat{\sigma}_X^2$ and $\hat{\sigma}_Y^2$, respectively. The distribution of T_n is now approximated by the distribution of

$$T_n^* = \frac{(\hat{\mu}_X^* - \hat{\mu}_Y^*) - (\hat{\mu}_X - \hat{\mu}_Y)}{\sqrt{\dfrac{\hat{\sigma}_X^{*2}}{n-1} + \dfrac{\hat{\sigma}_Y^{*2}}{m-1}}} \tag{3.14}$$

conditional on $\mathcal{X}$ and $\mathcal{Y}$, and is used to test $\mathsf{H} : \mu_X = \mu_Y$. This procedure is illustrated in Table 3.2. Some simulation results can be found in (Zoubir, 2001).

Table 3.2. *The bootstrap principle for testing the hypothesis* $\mathsf{H} : \mu_X = \mu_Y$
against $\mathsf{K} : \mu_X \neq \mu_Y$.

Step 0. *Experiment.* Conduct the experiment and collect the random data into two samples $\mathcal{X} = \{X_1, \ldots, X_n\}$ and $\mathcal{Y} = \{Y_1, \ldots, Y_m\}$.

Step 1. Calculate the statistic

$$T_n = \frac{\hat{\mu}_X - \hat{\mu}_Y}{\sqrt{\frac{\hat{\sigma}_X^2}{n-1} + \frac{\hat{\sigma}_Y^2}{m-1}}}.$$

Step 2. *Resampling.* Using a pseudo-random number generator, draw a random sample $\mathcal{X}^*$ of the same size as $\mathcal{X}$, with replacement, from $\mathcal{X}$ and a random sample $\mathcal{Y}^*$ of the same size as $\mathcal{Y}$, with replacement, from $\mathcal{Y}$.

Step 3. *Calculation of the bootstrap statistic.* From $\mathcal{X}^*$ and $\mathcal{Y}^*$, calculate

$$T_n^* = \frac{(\hat{\mu}_X^* - \hat{\mu}_Y^*) - (\hat{\mu}_X - \hat{\mu}_Y)}{\sqrt{\frac{\hat{\sigma}_X^{*2}}{n-1} + \frac{\hat{\sigma}_Y^{*2}}{m-1}}}.$$

Step 4. *Repetition.* Repeat Steps 2 and 3 many times to obtain a total of B bootstrap statistics $T_{n,1}^*, T_{n,2}^*, \ldots, T_{n,B}^*$.

Step 5. *Ranking.* Rank the collection $T_{n,1}^*, T_{n,2}^*, \ldots, T_{n,B}^*$ into increasing order to obtain $T_{n,(1)}^* \leq T_{n,(2)}^* \leq \cdots \leq T_{n,(B)}^*$.

Step 6. *Test.* A bootstrap test has the following form: reject H if $T_n < T_{(q_1)}^*$ or $T_n > T_{(q_2)}^*$, where $q_1 = \lfloor B\alpha/2 \rfloor$ and $q_2 = B - q_1 + 1$ is determined by the nominal level of significance of the test α.

Example 3.4.2 Testing the third-order cumulant

Let $\mathcal{X} = \{X_1, X_2, \ldots, X_n\}$ be a random sample from an unspecified distribution F. Define the third-order unbiased sample cumulant of X, computed from $\mathcal{X}$ as

$$\hat{c}_3 = \frac{1}{n(n-1)(n-2)} \left[n^2 \sum_{i=1}^{n} X_i^3 - 3n \sum_{i=1}^{n} X_i^2 \sum_{i=1}^{n} X_i + 2 \left(\sum_{i=1}^{n} X_i \right)^3 \right]$$

Let us construct a test to ascertain whether the third-order cumulant is zero, i.e.

$$\mathsf{H} : c_3 = 0$$

against a double-sided alternative

$$K : c_3 \neq 0.$$

Such a test can be considered as the second step for testing Gaussianity of data following Gasser (1975), see also Iskander *et al.* (1995).† Consider now the following three tests that can be used for this task:

(i) non-pivoted bootstrap test,
(ii) pivoted bootstrap test with a nested bootstrap (double bootstrap), and
(iii) pivoted bootstrap test with variance stabilisation.

In the non-pivoted bootstrap test we will use the following test statistic

$$T_n = |\hat{c}_3 - 0|$$

with its bootstrap equivalent

$$T_n^* = |\hat{c}_3^* - \hat{c}_3|.$$

In Section A1.6 of Appendix 1, we give a MATLAB code for the non-pivoted bootstrap test embedded in a Monte Carlo simulation to show that the test maintains the preset level. The number of bootstrap resamples in this case was set to $B = 1000$. In most practical applications the number of resamples should be of order $B \geq 10/\alpha$. For more guidelines on choosing B see the work of Hall and Titterington (1989). The simulation of a non-pivotal bootstrap test is also repeated in Section A1.6, using the function boottestnp.m from Appendix 2. This general function works with an arbitrary statistic.

Consider now the second case in which we will use the following test statistic

$$T_n = \frac{|\hat{c}_3 - 0|}{\hat{\sigma}}$$

and its bootstrap equivalent

$$T_n^* = \frac{|\hat{c}_3^* - \hat{c}_3|}{\hat{\sigma}^*}.$$

As mentioned earlier, the scaling parameters $\hat{\sigma}$ and $\hat{\sigma}^*$ are used so that the test statistics are asymptotically pivotal. In this case, evaluation of the scaling parameters will be performed with a double bootstrap procedure. In Section A1.6 of Appendix 1, we show the corresponding MATLAB code with $B_1 = 200$ and $B_2 = 25$. Note that the number of bootstrap resamples

---

† The first step of this test for Gaussianity is to test whether the kurtosis is zero.

is $B_1 \times B_2$ (see Section 3.4), because for each bootstrap sample $\mathcal{X}_b^*$, $b = 1, \ldots, B_1$, we draw resamples $\mathcal{X}_b^{**}$, $b = 1, \ldots, B_2$. Although the nested bootstrap brings a significant additional computational cost, we found that pivoted bootstrap hypothesis testing as well as pivoted bootstrap based confidence interval estimation are more accurate than the unpivoted ones.

As in the previous case we can use a general testing function boottest.m that is included in the Appendix 2. The above simulation is repeated in Section A1.6 of Appendix 1 using this function.

The last case in this example is a bootstrap hypothesis test with variance stabilisation. In Section A1.6 of Appendix 1, we omit the detailed code and use the function boottestvs.m from Appendix 2 instead.

Variance stabilisation has a great advantage in computational saving. The number of bootstrap resamples in this case is $B_1 \cdot B_2 + B_3$. The other advantage is that the parameters $B_1 = 100$ and $B_2 = 25$ used for variance estimation are sufficiently high in most practical cases while the number of resamples B_2 in the nested bootstrap may need to be increased in some applications (Tibshirani, 1988).

For all cases considered we have observed that bootstrap hypothesis tests maintain the preset level of significance α. The power of each of these bootstrap tests will depend on the alternatives and the number of samples available. For more details on bootstrap goodness-of-fit testing the reader is referred to the work of Zoubir and Iskander (1999).

3.5 Detection through regression

A pivotal statistic may not always be available. This makes the above techniques inapplicable. An alternative in this case is to resample under the null hypothesis (Tibshirani, 1992). The idea of such an approach is not entirely new and can be related to bootstrap distance tests (Beran, 1986; Hinkley, 1988; Romano, 1988). A method based on this idea has been proposed by Ong (2000) for detection when a parametric model for the signal applies. Let the observations be

$$X_t = s_t(\theta, \boldsymbol{\vartheta}) + Z_t, \qquad t = 1, \ldots, n, \tag{3.15}$$

where θ is the parameter to be tested and $\boldsymbol{\vartheta} = (\vartheta_1, \ldots, \vartheta_p)'$ are other unknown parameters called *nuisance parameters*. The functions $s_1, \ldots, s_n$ are assumed to be continuous, such as in the detection of a sinusoid with unknown amplitude, phase and frequency, $s_t(\theta, \boldsymbol{\vartheta}) = \theta \cos(\vartheta_1 t + \vartheta_2)$. The proposed procedure for testing the hypotheses $\mathsf{H} : \theta = \theta_0$ against $\mathsf{K} : \theta > \theta_0$

is given in Table 3.3. The test statistic is $T_n = \hat{\theta}$ and is compared with a threshold found from the bootstrap distribution of $T_n^* = \hat{\theta}^*$.

Table 3.3. *A regression based method for detection.*

Step 1. Fit a model, $X_t = s_t(\theta, \boldsymbol{\vartheta}) + Z_t$, to the data $X_1, \ldots, X_n$, and find the residuals,

$$r_t = X_t - s_t(\hat{\theta}, \hat{\boldsymbol{\vartheta}}), \qquad t = 1, \ldots, n, \qquad (3.16)$$

where $\hat{\theta}$ and $\hat{\boldsymbol{\vartheta}}$ are the parameter estimates of the fitted model found, for example, by least squares. Estimate the interference by the (sample mean) centred residuals, $\hat{Z}_t = r_t - \bar{r}$, $t = 1, \ldots, n$.

Step 2. Resample the centred residuals $\hat{Z}_1, \ldots, \hat{Z}_n$ to get $\boldsymbol{Z}^* = (Z_1^*, \ldots, Z_n^*)'$.

Step 3. Generate bootstrap data under the null hypothesis, i.e., $\theta = \theta_0$,

$$X_t^* = s_t(\theta_0, \hat{\boldsymbol{\vartheta}}) + Z_t^*, \qquad t = 1, \ldots, n. \qquad (3.17)$$

Step 4. Calculate the bootstrap statistic, $T_n^*(\boldsymbol{X}^*) = \hat{\theta}^*$ in the same way $\hat{\theta}$ was found, but using X_t^*, $t = 1, \ldots, n$ instead.

Step 5. Repeat Steps 2–4 many times to get B bootstrap statistics, $T_{n,1}^*, \ldots, T_{n,B}^*$.

Step 6. Sort the bootstrap statistics: $T_{n,(1)}^* \leq \cdots \leq T_{n,(B)}^*$.

Step 7. Reject $\mathsf{H} : \theta = \theta_0$ and decide a signal is present if $T_n > T_{(q)}^*$ where $T_n = \hat{\theta}$ and $q = \lceil (B+1)(1-\alpha) \rceil$.

The procedure is more complicated than the pivot method and requires a parametric model so that the bootstrap data can be constructed under the null hypothesis. However, it does not need a pivot to work.

Note that the pivot method can still be applied using a parametric model but instead of generating data under the null, the estimated parameter $\hat{\theta}$ is put back in. Table 3.4 gives the modified procedure with the changes in Steps 3, 4 and 7.

The regression method is not designed only to overcome problems with the pivot. The method is used in many practical detection problems as well as bootstrap modelling (see Chapter 4). In particular, we will use this regression method for a bootstrap matched filter developed in Section 3.6.

We will summarise this section with a signal processing example in which bootstrap based hypothesis testing is successfully employed. Some aspects of this example have been covered in a bootstrap tutorial paper by Zoubir and Boashash (1998).

Table 3.4. *The pivot method using a parametric model.*

Step 1. Fit a model, $X_t = s_t(\theta, \boldsymbol{\vartheta}) + Z_t$, to the data $X_1, \ldots, X_n$, and find the residuals,

$$r_t = X_t - s_t(\hat{\theta}, \hat{\boldsymbol{\vartheta}}), \qquad t = 1, \ldots, n, \tag{3.18}$$

where $\hat{\theta}$ and $\hat{\boldsymbol{\vartheta}}$ are the parameter estimates of the fitted model found, for example, by least squares. Estimate the interference by the centred residuals, $\hat{Z}_t = r_t - \bar{r}$, $t = 1, \ldots, n$.

Step 2. Resample the centred residuals $\hat{Z}_1, \ldots, \hat{Z}_n$ to get $\boldsymbol{Z}^* = (Z_1^*, \ldots, Z_n^*)'$.

Step 3. Generate bootstrap data,

$$X_t^* = s_t(\hat{\theta}, \hat{\boldsymbol{\vartheta}}) + Z_t^*, \qquad t = 1, \ldots, n. \tag{3.19}$$

Step 4. Calculate bootstrap statistic, $T^*(\boldsymbol{X}^*) = (\hat{\theta}(\boldsymbol{X}^*) - \hat{\theta})/\hat{\sigma}_{\hat{\theta}^*}$.

Step 5. Repeat Steps 2–4 many times to get B bootstrap statistics, $T_1^*, \ldots, T_B^*$.

Step 6. Sort the bootstrap statistics: $T_{(1)}^* \leq \cdots \leq T_{(B)}^*$.

Step 7. Reject $\mathsf{H} : \theta = \theta_0$ and decide a signal is present if $T > T_{(q)}^*$ where $T = (\hat{\theta} - \theta_0)/\hat{\sigma}_{\hat{\theta}}$ and $q = \lceil (B+1)(1-\alpha) \rceil$.

Example 3.5.1 MISO linear system analysis

Consider a signal processing application in which we have a multiple-input single-output (MISO) linear system as illustrated in Figure 3.5. Herein, an r vector-valued stationary process $\boldsymbol{S}_t = (S_{1,t}, \ldots, S_{r,t})'$ is transmitted through a linear time-invariant system whose r vector-valued impulse response is $\boldsymbol{g}_t = (g_{1,t}, \ldots, g_{r,t})'$. We will assume that the linear system is stable. The system output is buried in a stationary zero-mean noise process $\mathcal{E}_t$ and received as a stationary process Z_t, where $\mathcal{E}_t$ and $\boldsymbol{S}_t$ are assumed to be independent for any $t = 0, \pm 1, \pm 2, \ldots$ For such a model, we can write

$$Z_t = \sum_{u=-\infty}^{\infty} \boldsymbol{g}_u' \boldsymbol{S}_{t-u} + \mathcal{E}_t. \tag{3.20}$$

Let

$$\boldsymbol{G}(\omega) = (G_1(\omega), \ldots, G_r(\omega))' = \sum_{t=-\infty}^{\infty} \boldsymbol{g}_t e^{-j\omega t}$$

be the unknown transfer function of the system.

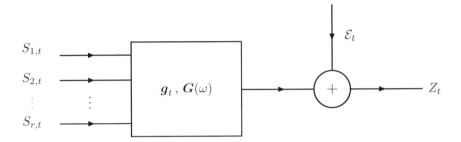

Fig. 3.5. A typical multiple input single output system (MISO).

The question we wish to answer is: *Which element $G_l(\omega)$, $1 \le l \le r$, is zero at a given frequency ω?* Under H, this would imply that Z_t does not contain any information at ω, contributed by the lth signal component of S_t, $S_{l,t}$, $1 \le l \le r$. This situation occurs in many applications where one is interested in approximating a vector-valued time series by a version of itself plus noise, but restraining the new series to be of reduced dimension (in this case a scalar). Then, the problem is to detect channels (frequency responses) that have bandstop behaviour at certain frequencies.

A specific example is a situation where one is interested in finding suitable vibration sensor positions to detect tool wear or a break in a milling process. One would distribute sensors on the spindle fixture and one sensor as a reference on the work piece, which would not be accessible in a serial production. Based on observations of the vibration sensors, one would decide the suitability of a sensor position on the spindle fixture based on the structural behaviour of the fixture at some given frequencies, which would have been assumed to be linear and time invariant. This problem is currently resolved using heuristic arguments. A similar problem in motor engines is discussed by Zoubir and Böhme (1995).

Another application in vertical seismic profiling requires a detailed knowledge of the filter constituted by the various layers constituting the vertical earth profile at a particular point in space. In such an application, waves are emitted in the ground which propagate through and reflect on the diopter which separate the various layers characterised by different acoustic impedances. The transmission function for the range of relevant frequencies is of crucial importance, as the filter characteristics vary from location to location. Knowledge of the frequency transfer function of the various earth layers filter contributes to a proper modelling of the earth surface, and then to a decision as to the likelihood of the presence of gas and/or oil in a particular place.

The considered problem is by no means limited to these two practical examples. Existence of such a MISO system problem can be encountered in other engineering applications including smart antenna systems (Ruly *et al.*, 2001) and control (Malti *et al.*, 1998; Besson and Shenton, 2000).

Returning to our problem, we would like to test whether $G_l(\omega) = 0$, $1 \leq l \leq r$, that is, we test the hypothesis

$$\mathsf{H} : G_l(\omega) = 0$$

with unspecified $\boldsymbol{G}^{(l)}(\omega)$ against the two-sided alternative

$$\mathsf{K} : G_l(\omega) \neq 0 \,,$$

where $\boldsymbol{G}^{(l)}(\omega)$ is such that $\boldsymbol{G}(\omega) = (\boldsymbol{G}^{(l)}(\omega)', G_l(\omega))'$, with $G_l(\omega)$ being an arbitrary frequency response that represents the transfer function of the filter that transforms the signal $S_{l,t}$ by time-invariant and linear operations, and $\boldsymbol{G}^{(l)}(\omega) = (G_1(\omega), G_2(\omega), \ldots, G_{l-1}(\omega), G_{l+1}(\omega), \ldots, G_r(\omega))'$ is the vector of transfer functions obtained from $\boldsymbol{G}(\omega)$ by deleting the component $G_l(\omega)$.

Let $\boldsymbol{S}_t$ and Z_t be given for n independent observations of length T each. By taking the finite Fourier transform of both sides of (3.20), we obtain (omitting the error term† $o_{a.s.}(1)$) the complex regression

$$\boldsymbol{d}_Z(\omega) = \boldsymbol{d}_{\boldsymbol{s}}(\omega)\boldsymbol{G}(\omega) + \boldsymbol{d}_{\mathcal{E}}(\omega) \,, \qquad (3.21)$$

where $\boldsymbol{d}_{\boldsymbol{s}}(\omega) = (\boldsymbol{d}_{S_.}(\omega), \ldots, \boldsymbol{d}_{S_r}(\omega))$, $\boldsymbol{d}_{S_l}(\omega) = (d_{S_l}(\omega, 1), \ldots, d_{S_l}(\omega, n))'$, $l = 1, \ldots, r$, $\boldsymbol{d}_Z(\omega) = (d_Z(\omega, 1), \ldots, d_Z(\omega, n))'$, and

$$d_Z(\omega, i) = \sum_{t=0}^{T-1} w(t/T) \cdot Z_{t,i} \, e^{-j\omega t}, \qquad i = 1, \ldots, n, \qquad (3.22)$$

is the normalised finite Fourier transform of the ith data block $Z_{t,i}$, $i = 1, \ldots, n$, of Z_t and $w(u), u \in \mathbb{R}$, is a smooth window that vanishes outside the interval $[0, 1]$.

Based on the observations of $\boldsymbol{S}_t$ and Z_t, for $t = 0, 1, \ldots, T - 1$, and $i = 1, \ldots, n$, we first calculate the least-squares estimate of $\boldsymbol{G}(\omega)$,

$$\begin{aligned} \hat{\boldsymbol{G}}(\omega) &= (\boldsymbol{d}_{\boldsymbol{s}}(\omega)^H \boldsymbol{d}_{\boldsymbol{s}}(\omega))^{-1}(\boldsymbol{d}_{\boldsymbol{s}}(\omega)^H \boldsymbol{d}_Z(\omega)) \qquad (3.23) \\ &= \hat{\boldsymbol{C}}_{SS}(\omega)^{-1}\hat{\boldsymbol{C}}_{ZS}(\omega)' \,, \end{aligned}$$

where $\hat{\boldsymbol{C}}_{SS}(\omega)$ and $\hat{\boldsymbol{C}}_{ZS}(\omega)$ are spectral densities obtained by averaging the

---

† $o_{a.s.}(1)$ is an error term that tends to zero almost surely as $T \to \infty$ (Brillinger, 1981).

corresponding periodograms over n independent data records, and ^{H} denotes the Hermitian operation.

Conventional techniques assume the number of observations T to be large so that the finite Fourier transform $d_{\mathcal{E}}(\omega)$ becomes asymptotically complex Gaussian (Brillinger, 1981). Under this assumption and H, the statistic

$$\hat{\theta}(\omega) = \frac{\parallel d_Z(\omega) - d_{s \cdot l \cdot}(\omega)\hat{G}^{(l)}(\omega) \parallel^2 - \parallel d_Z(\omega) - d_s(\omega)\hat{G}(\omega) \parallel^2}{\parallel d_Z(\omega) - d_s(\omega)\hat{G}(\omega) \parallel^2 /(n-r)} \qquad (3.24)$$

is assumed to be F-distributed with 2 and $2(n-r)$ degrees of freedom. Note that if χ_1^2 and χ_2^2 are independent random variables having chi-square distributions of ν_1 and ν_2 degrees of freedom, respectively, then $(\chi_1^2/\nu_1)/(\chi_2^2/\nu_2)$ has an F-distribution with ν_1 and ν_2 degrees of freedom. In Equation (3.24), $d_{s \cdot l \cdot}(\omega)$ is defined as

$$d_{s \cdot l \cdot}(\omega) = (d_{S \cdot}(\omega), \ldots, d_{S_{l-}}(\omega), d_{S_l \cdot \cdot}(\omega), \ldots, d_{S_r}(\omega))'$$

and is obtained from $d_s(\omega) = (d_{s \cdot l \cdot}(\omega), d_{S_l}(\omega))$ by deleting the lth vector $d_{S_l}(\omega)$, and $\hat{G}^{(l)}(\omega) = (d_{s \cdot l \cdot}(\omega)^H d_{s \cdot l \cdot}(\omega))^{-1}(d_{s \cdot l \cdot}(\omega)^H d_Z(\omega))$ (Shumway, 1983).

The hypothesis H is rejected at a level α if the statistic (3.24) exceeds the $100(1-\alpha)\%$ quantile of the F-distribution.

The use of the F-distribution in the case where $d_{\mathcal{E}}(\omega)$ is non-Gaussian is not valid. To find the distribution of the statistic (3.24) in the more general case, we will use a procedure based on the bootstrap, described in Table 3.5.

Note that several regression models are available and depending upon which model the analysis is based, we have a different resampling procedure (Hall, 1992). In the procedure described in Table 3.5, we assumed the pairs $(d_S(\omega, i), d_{\mathcal{E}}(\omega, i))$ to be independent and identically distributed, with $d_S(\omega, i)$ and $d_{\mathcal{E}}(\omega, i)$ independent.

An alternative bootstrap approach to the one described in Table 3.5 is based on the fact that (3.24) can be written as

$$\hat{\theta}(\omega) = (n-r)\frac{|\hat{R}_{ZS}(\omega)|^2 - |\hat{R}_{ZS \cdot l \cdot}(\omega)|^2}{1 - |\hat{R}_{ZS}(\omega)|^2}, \qquad (3.25)$$

where

$$|\hat{R}_{ZS}(\omega)|^2 = \frac{\hat{C}_{ZS}(\omega)\hat{C}_{SS}(\omega)^{-1}\hat{C}_{SZ}(\omega)}{\hat{C}_{ZZ}(\omega)} \qquad (3.26)$$

and

$$|\hat{R}_{ZS \cdot l \cdot}(\omega)|^2 = \frac{\hat{C}_{ZS \cdot l \cdot}(\omega)\hat{C}_{S \cdot l \cdot S \cdot l \cdot}(\omega)^{-1}\hat{C}_{S \cdot l \cdot Z}(\omega)}{\hat{C}_{ZZ}(\omega)} \qquad (3.27)$$

Table 3.5. *The bootstrap principle for the example of regression analysis in a MISO system.*

Step 0. *Experiment.* Conduct the experiment and calculate the frequency data $\boldsymbol{d}_S(\omega, 1), \ldots, \boldsymbol{d}_S(\omega, n)$, and $d_Z(\omega, 1), \ldots, d_Z(\omega, n)$.

Step 1. *Resampling.* Conduct two totally independent resampling operations in which a random sample $\{\boldsymbol{d}_S^*(\omega, 1), \ldots, \boldsymbol{d}_S^*(\omega, n)\}$ is drawn, with replacement, from $\{\boldsymbol{d}_S(\omega, 1), \ldots, \boldsymbol{d}_S(\omega, n)\}$, where

$$\boldsymbol{d}_S(\omega, i) = (d_{S_1}(\omega, i), \ldots, d_{S_r}(\omega, i)),$$

$i = 1, \ldots, n$, and a resample $\{d_{\hat{\varepsilon}}^*(\omega, 1), \ldots, d_{\hat{\varepsilon}}^*(\omega, n)\}$ is drawn, with replacement, from the random sample $\{d_{\hat{\varepsilon}}(\omega, 1), \ldots, d_{\hat{\varepsilon}}(\omega, n)\}$, collected into the vector $\boldsymbol{d}_{\hat{\varepsilon}}(\omega) = (d_{\hat{\varepsilon}}(\omega, 1), \ldots, d_{\hat{\varepsilon}}(\omega, n))'$, so that

$$\boldsymbol{d}_{\hat{\varepsilon}}(\omega) = \boldsymbol{d}_Z(\omega) - \boldsymbol{d}_s(\omega)\hat{\boldsymbol{G}}(\omega)$$

are the residuals of the regression model (3.21).

Step 2. *Generation of bootstrap data.* Centre the frequency data resamples and compute

$$\boldsymbol{d}_Z^*(\omega) = \boldsymbol{d}_s^*(\omega)\hat{\boldsymbol{G}}(\omega) + \boldsymbol{d}_{\hat{\varepsilon}}^*(\omega).$$

The joint distribution of $\{(\boldsymbol{d}_S^*(\omega, i), d_Z^*(\omega, i)), 1 \leq i \leq n\}$, conditional on

$$\mathcal{X}(\omega) = \{(\boldsymbol{d}_S(\omega, 1), d_Z(\omega, 1)), \ldots, (\boldsymbol{d}_S(\omega, n), d_Z(\omega, n))\}$$

is the bootstrap estimate of the unconditional joint distribution of $\mathcal{X}(\omega)$.

Step 3. *Calculation of bootstrap estimates.* With the new $\boldsymbol{d}_Z^*(\omega)$ and $\boldsymbol{d}_s^*(\omega)$, calculate $\hat{\boldsymbol{G}}^*(\omega)$, using (3.23) but with the resamples $\boldsymbol{d}_Z^*(\omega)$ and $\boldsymbol{d}_s^*(\omega)$, replacing $\boldsymbol{d}_Z(\omega)$ and $\boldsymbol{d}_s(\omega)$, respectively.

Step 4. *Calculation of the bootstrap statistic.* Calculate the statistic given in (3.24), replacing $\boldsymbol{d}_s(\omega)$, $\boldsymbol{d}_{s^{(l)}}(\omega)$, $\boldsymbol{d}_Z(\omega)$, $\hat{\boldsymbol{G}}(\omega)$, and $\hat{\boldsymbol{G}}^{(l)}(\omega)$ by their bootstrap counterparts to yield $\hat{\theta}^*(\omega)$.

Step 5. *Repetition.* Repeat Steps 1–4 a large number of times, say B, to obtain $\hat{\theta}_1^*(\omega), \ldots, \hat{\theta}_B^*(\omega)$.

Step 6. *Distribution estimation.* Approximate the distribution of $\hat{\theta}(\omega)$, given in (3.24), by the obtained distribution of $\hat{\theta}^*(\omega)$.

---

are respectively the sample multiple coherence of Z_t with $\boldsymbol{S}_t$ and Z_t with

$$\boldsymbol{S}_t^{(l)} = (S_{1,t}, S_{2,t}, \ldots, S_{l-1,t}, S_{l+1,t}, \ldots, S_{r,t})'$$

at frequency ω. Herein, the spectral densities

$$\hat{\boldsymbol{C}}_{ZS}(\omega), \hat{\boldsymbol{C}}_{ZS\cdot l\cdot}(\omega), \hat{\boldsymbol{C}}_{SS}(\omega), \hat{\boldsymbol{C}}_{S\cdot l\cdot S\cdot l\cdot}(\omega) \quad \text{and} \quad \hat{\boldsymbol{C}}_{ZZ}(\omega)$$

in (3.26) and (3.27) are obtained by averaging periodograms of n independent data records.

Alternatively to Table 3.5, we could proceed as described in Table 3.6 to estimate the distribution of $\hat{\theta}(\omega)$. The main difference with this approach compared to the previous one is that the resampling procedure does not take into consideration the assumed complex regression model (3.21).

Table 3.6. *An alternative bootstrap approach for the example of regression analysis in a MISO system.*

Step 0. *Experiment.* Conduct the experiment and calculate the frequency data $\boldsymbol{d}_S(\omega, 1), \ldots, \boldsymbol{d}_S(\omega, n)$, and $d_Z(\omega, 1), \ldots, d_Z(\omega, n)$.

Step 1. *Resampling.* Using a pseudo-random number generator, draw a random sample $\mathcal{X}(\omega)^*$ (of the same size), with replacement, from

$$\mathcal{X}(\omega) = \{(\boldsymbol{d}_S(\omega, 1), d_Z(\omega, 1)), \ldots, (\boldsymbol{d}_S(\omega, n), d_Z(\omega, n))\}.$$

Step 2. *Calculation of the bootstrap statistic.* From $\mathcal{X}(\omega)^*$, calculate $\hat{\theta}^*(\omega)$, the bootstrap analogue of $\hat{\theta}(\omega)$ given by (3.25).

Step 3. *Repetition.* Repeat Steps 1 and 2 many times to obtain a total of B bootstrap statistics $\hat{\theta}_1^*(\omega), \ldots, \hat{\theta}_B^*(\omega)$.

Step 4. *Distribution estimation.* Approximate the distribution of $\hat{\theta}(\omega)$, given in (3.25), by the so obtained bootstrap distribution.

It is worthwhile emphasising that in practice bootstrap resampling should be performed to reflect the model characteristics. If we assume that the data is generated from model (3.21), we should use the method given in Table 3.5 to estimate the distribution of the test statistic. Resampling from $\mathcal{X}(\omega)$ will not necessarily generate data satisfying the *assumed* model. In our application, the regression model (3.21) is an approximation only and its validity is questionable in the case where the number of observations is small. Notice that Equation (3.25) is a measure (see the work of Zoubir and Böhme (1995)) of the extent to which the signal $S_{l,t}$ contributes in Z_t and can be derived heuristically, without use of (3.24), that is based on regression (3.21).

Comparative studies of the two indicated resampling methods will show that bootstrapping coherences as discussed in Table 3.6 gives similar test results as the approach in Table 3.5.

Numerical results We simulate $n = 20$ independent records of a vector-valued signal $\boldsymbol{S}_t$ with $r = 5$. The model used to generate a component $S_{l,t}$, $l = 1, \ldots, 5$ is as follows:

$$S_{l,t} = \sum_{k=1}^{K} A_{k,l} \cos(\omega_k t + \Phi_{k,l}) + U_{l,t}, \quad l = 1, \ldots, 5. \quad (3.28)$$

Herein, $A_{k,l}$ and $\Phi_{k,l}$ are mutually independent random amplitudes and phases, respectively, ω_k are arbitrary resonance frequencies for $k = 1, \ldots, K$ and $U_{l,t}$ is a white noise process, $l = 1, \ldots, r$. We fix $K = 4$ and generate records of length $T = 128$ each, using a uniform distribution for both the phase and the amplitude on the interval $[0, 2\pi)$ and $[0, 1)$, respectively. Let the resonance frequencies be $f_1 = 0.1, f_2 = 0.2, f_3 = 0.3$ and $f_4 = 0.4$, all normalised, where $f_k = \omega_k / 2\pi$, $k = 1, \ldots, 4$, and add uniformly distributed noise U_t to the generated signal. A typical spectrum of $S_{l,t}$, $l = 1, \ldots, r$ obtained by averaging 20 periodograms is depicted in Figure 3.6.

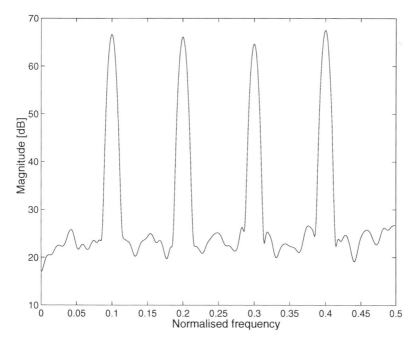

Fig. 3.6. Spectrum of $S_t^{(l)}$ defined in (3.28), obtained by averaging 20 periodograms.

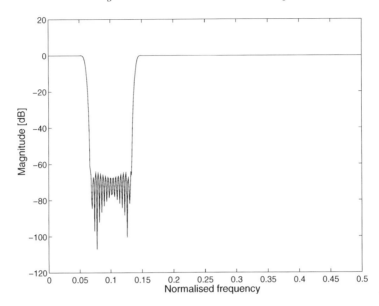

Fig. 3.7. Frequency response of the first channel, $G_1(\omega)$, obtained using an FIR filter with 256 coefficients.

We then generate bandstop FIR filters (with 256 coefficients) with bands centred about the four resonance frequencies f_1, f_2, f_3 and f_4. As an example, the frequency response of the first filter, $G_1(\omega)$, is given in Figure 3.7.

We filter $\boldsymbol{S}_t$ and add independent uniformly distributed noise $\mathcal{E}_t$ to the filtered signal to generate Z_t. The SNR of the output signal is 5 dB with respect to the component $S_{l,t}$ with highest power. A plot of a spectral estimate of Z_t, obtained by averaging 20 periodograms, is given in Figure 3.8.

We arbitrarily select one transfer function $G_l(\omega)$, $l = 1, \ldots, r$, and test H. Using the procedure of Table 3.1 with 1000 and 30 bootstrap resamples for a quantile and variance estimate, respectively, we reject H at a level of significance of 5% if ω does not fall in the stopband. Otherwise, we retain H. In the simulations we have performed, with both methods described in Table 3.5 and Table 3.6, the level of significance obtained was never above the nominal value.

Figures 3.9 and 3.10 show the bootstrap distribution of the statistic $(\hat{\theta}^*(\omega) - \hat{\theta}(\omega))/\hat{\sigma}^*(\omega)$, using the procedures described in Table 3.5 and Table 3.6, where $\hat{\theta}(\omega)$ is given by (3.24) and (3.25), respectively. We test $\mathsf{H}_2 : G_2(\omega) = 0$, against the two-sided alternative, where ω was taken to be the centre frequency of the bandstop. With both methods, the hypothesis is retained.

These numerical examples show that it is possible to perform tests in complicated situations but with minimal assumptions. In this situation, classical

(F-) tests are only valid if the number of observations T is assumed to be large and asymptotic results for the finite Fourier transform of stationary data hold (Brillinger, 1983).

Note that in practice one is interested in performing the above discussed tests not only at one frequency but at a set of multiple frequencies simultaneously. A discussion of bootstrap procedures to perform multiple tests can be found in the work of Zoubir (1994). A practical application of the method described in Table 3.6 was discussed by Zoubir and Böhme (1995).

3.6 The bootstrap matched filter

Recall the basic detection problem from Figure 3.2 where the observations

$$X = \theta s + Z$$

of a transmitted signal $s = (s_1, \ldots, s_n)'$ embedded in noise Z are collected at the receiver. The detection is performed by finding a suitable function $T_n = T_n(x)$ used as a test statistic and compared against a threshold T_α.

If the interfering noise is zero-mean white Gaussian with a known variance σ^2 then a simple solution to the problem exists (Neyman–Pearson lemma).

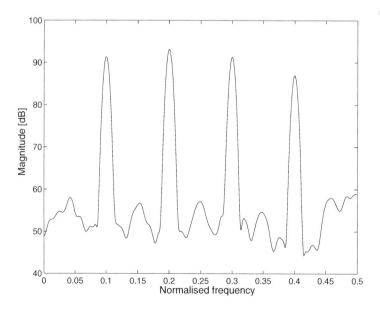

Fig. 3.8. Spectrum of Z_t, using an average of 20 periodograms.

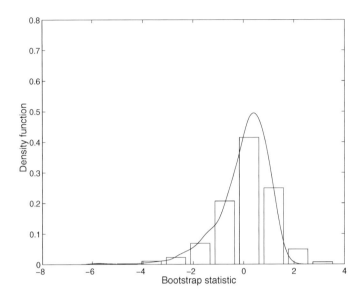

Fig. 3.9. Histogram of 1000 bootstrap values of the statistic $(\hat{\theta}^*(\omega) - \hat{\theta}(\omega))/\hat{\sigma}^*(\omega)$, where $\hat{\theta}(\omega)$ is given in (3.24), at a frequency bin where the hypothesis $H_2: G_2(\omega) = 0$ is retained.

We define

$$T_n = \frac{s'x}{\sigma\sqrt{s's}}$$

and find the threshold to be $T_\alpha = \Phi^{-1}(1 - \alpha)$, where α is the set level of significance.

If the noise variance is unknown, the constant false alarm rate (CFAR) matched filter is the optimal solution to the detection problem in the sense that it is uniformly most powerful within the class of detectors that are invariant to a scaling transformation of the observations (Scharf, 1991). That is, it is the detector with the highest P_D whose performance in terms of P_F and P_D remains the same regardless of the observations being scaled by an arbitrary constant. The nominal level is maintained by the CFAR matched filter as a direct result of its scale-invariant property. The test statistic and threshold are respectively,

$$T = s'X/(\hat{\sigma}\sqrt{s's}),$$

and

$$T_\alpha = t_{n-1,\alpha},$$

where $t_{n-1,\alpha}$ denotes the $100(1 - \alpha)$th percentile of the t-distribution with

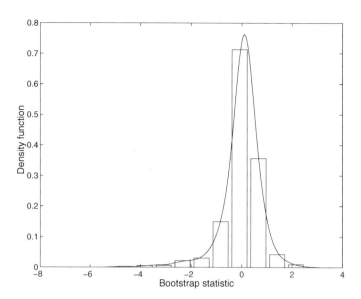

Fig. 3.10. Histogram of 1000 bootstrap values of the statistic $(\hat{\theta}^*(\omega) - \hat{\theta}(\omega))/\hat{\sigma}^*(\omega)$, where $\hat{\theta}(\omega)$ is given in (3.25), at a frequency bin where the hypothesis $H_2 : G_2(\omega) = 0$ is retained.

$n - 1$ degrees of freedom, and $\hat{\sigma}^2$ is given by

$$\hat{\sigma}^2 = \frac{\left\| X - s \frac{s'X}{s's} \right\|^2}{n - 1},$$

where $\|x\| = \sqrt{x'x}$.

The performance of the CFAR matched filter is illustrated in Figures 3.11 and 3.12 (Scharf, 1991).

The matched filter and the CFAR matched filter are designed to be optimal for Gaussian interference. Although they also show a high probability of detection in some non-Gaussian cases, they are unable to maintain the preset level of significance for small sample sizes. Thus we can identify two cases where the matched filter fails:

- where the interference/noise is non-Gaussian
- where the data size is small.

For the particular non-Gaussian case, optimal signal detection techniques are available (Conte *et al.*, 1996), although their power against traditional CFAR detection is evident only for small numbers of pulses (Iskander, 1998).

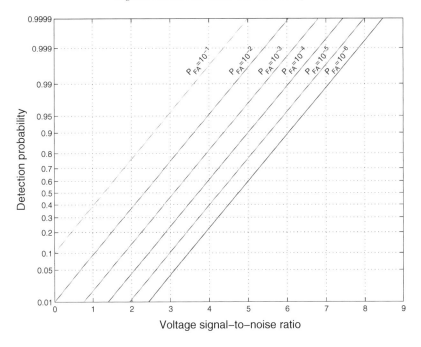

Fig. 3.11. Probability of detection of a CFAR matched filter for different levels of probability of false alarm.

Thus, the goal is to develop techniques which require little in the way of modelling and assumptions.

Since any detection problem can be put in the framework of hypothesis testing and since we have argued that bootstrap based techniques are a powerful tool, it is expected that a bootstrap based solution to the two problems mentioned above exists. The bootstrap is also attractive when detection is associated with estimation, e.g., estimation of the number of signals. This and other applications will be discussed later.

Let us recall the basic bootstrap signal detection algorithm of Figure 3.4 and Table 3.1. We will now proceed with a bootstrap matched filter algorithm following similar steps as in Table 3.1 and invoking the concept of regression based detection (see Section 3.5). The algorithm for a bootstrap matched filter is shown in Table 3.7.

The difference between this algorithm and the general procedure from Figure 3.4 is only in the first two steps. That is, we are required to calculate the estimator of θ and then form a linear regression to obtain a set of residuals w_t, $t = 1, \ldots, n$.

The bootstrap matched filter is consistent, i.e., $P_d \to 1$ as $n \to \infty$. Ong

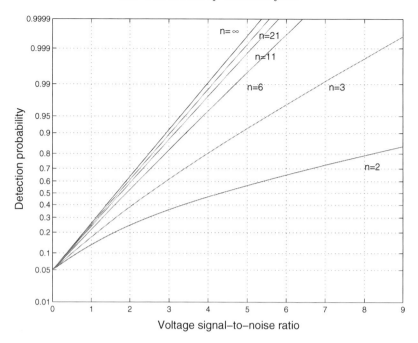

Fig. 3.12. Probability of detection of a CFAR matched filter for a different number of samples.

and Zoubir (2000a) have shown that for $s_t < \infty$, $t = 1, \ldots, n$, the probability of false alarm for the bootstrap matched filter is

$$\mathsf{P}_F^* = \alpha + O\left(n^{-1}\right).$$

Example 3.6.1 Performance of the matched filter: Gaussian noise

Consider a sinusoidal signal $s_t = A\sin(2\pi(t-1)/6)$, $t = 1, \ldots, n$, $n = 10$, buried in white zero-mean Gaussian noise with unit variance. We will evaluate the performance of the matched filter and the CFAR matched filter against their two bootstrap counterparts. In the bootstrap, we set the number of resamples to $B_1 = 1000$ and use a nested bootstrap with $B_2 = 25$ to estimate the variance of the estimator as in the MATLAB code of Section A1.7 of Appendix 1. We run 100 Monte Carlo simulations and evaluate the probability of false alarm and the probability of detection. In Appendix 2 we included the function bootmf.m that outputs the detection result (1 or 0) for the four detector structures mentioned above. The results of these simulations are shown in Table 3.8.

Table 3.7. *The bootstrap matched filter.*

Step 0. *Experiment.* Run the experiment and collect the data x_t, $t = 1, \ldots, n$.

Step 1. *Estimation.* Compute the least squares estimate $\hat{\theta}$ of θ, $\hat{\sigma}_{\hat{\theta}}$, and

$$T_n = \left. \frac{\hat{\theta} - \theta}{\hat{\sigma}_{\hat{\theta}}} \right|_{\theta = 0}.$$

Step 2. *Resampling.* Compute the residuals

$$\hat{w}_t = x_t - \hat{\theta} s_t, \qquad t = 1, \ldots, n,$$

and after centring, resample the residuals, assumed to be iid, to obtain $\hat{w}_t^*$, $t = 1, \ldots, n$.

Step 3. *Bootstrap test statistic.* Compute new measurements

$$x_t^* = \hat{\theta} s_t + \hat{w}_t^*, \qquad t = 1, \ldots, n,$$

and the least squares estimate $\hat{\theta}^*$ based on the resamples x_t^*, $t = 1, \ldots, n$, and

$$T_n^* = \frac{\hat{\theta}^* - \hat{\theta}}{\hat{\sigma}_{\hat{\theta}^*}}.$$

Step 4. *Repetition.* Repeat Steps 2 and 3 a large number of times to obtain $T_{n,1}^*, \ldots, T_{n,B}^*$.

Step 5. *Bootstrap test.* Sort $T_{n,1}^*, \ldots, T_{n,B}^*$ to obtain $T_{n,(1)}^* \leq \cdots \leq T_{n,(B)}^*$. Reject H if $T_n \geq T_{n,(q)}^*$, where $q = \lfloor (1 - \alpha)(B + 1) \rfloor$.

Table 3.8. *Estimated probability of false alarm, $\hat{\mathsf{P}}_F$, and probability of detection, $\hat{\mathsf{P}}_D$ for a sinusoidal signal embedded in standard Gaussian white noise.*

Detector	$\hat{\mathsf{P}}_F$ [%]	$\hat{\mathsf{P}}_D$ [%]
Matched Filter (MF)	5	99
CFAR MF	5	98
Bootstrap (σ known)	4	99
Bootstrap (σ unknown)	5	98

Example 3.6.2 Performance of the matched filter: non-Gaussian noise

Consider the same sinusoidal signal as in Example 3.6.1. Instead of Gaussian noise, we now use a Gaussian mixture

$$W_t \sim \sum_{i=1}^{4} a_i \mathcal{N}(\mu_i, \sigma_i),$$

with $\boldsymbol{a} = (0.5, 0.1, 0.2, 0.2)$, $\boldsymbol{\mu} = (-0.1, 0.2, 0.5, 1)$, and $\boldsymbol{\sigma} = (0.25, 0.4, 0.9, 1.6)$. We set the SNR to -6 dB. The results are shown in Table 3.9.

Table 3.9. *Estimated probability of false alarm, $\hat{P}_F$, and probability of detection, $\hat{P}_D$ for a sinusoidal signal embedded in Gaussian mixture noise.*

Detector	$\hat{P}_F$ [%]	$\hat{P}_D$ [%]
Matched Filter (MF)	8	83
CFAR MF	7	77
Bootstrap (σ known)	5	77
Bootstrap (σ unknown)	7	79

It is important to note that the classical matched filter as well as the CFAR matched filter do not always maintain the preset level of significance when the interference is non-Gaussian. The results of a similar but much more extensive study have been reported by Ong and Zoubir (2000a).

3.6.1 Tolerance interval bootstrap matched filter

The thresholds calculated in the bootstrap detectors are random variables conditional on the data. Thus, it is of interest to keep them as steady as possible. Formally, we would like to set β to specify our confidence that the set level is maintained. To achieve this, the required threshold $T^*_{(r)}$ must fulfil

$$\mathsf{Prob}\left[\mathsf{Prob}\left[T_n > T^*_{(r)}|\mathsf{H}\right] \leq \alpha\right] \geq \beta$$

Finding threshold r can be achieved through the theory of tolerance intervals. Here, we make a short digression before answering the threshold problem. Consider the following problem:

Given an iid sample $X_1, \ldots, X_n$, find two order statistics, $X_{(r)}$ and $X_{(n-s+1)}$

such that $[X_{(r)},\ X_{(n-s+1)}]$ will encompass at least $100(1-\alpha)\%$ of the underlying distribution, F, with probability at least β,

$$\mathsf{Prob}\left[1-\alpha \leq \int_{X_{\cdot r \cdot}}^{X_{\cdot n-s \cdot\ \cdot\cdot}} dF\right] \geq \beta.$$

From the theory of distribution free tolerance intervals we must find r and s such that

$$1 - F_B(1-\alpha; n-r-s+1, r+s) \geq \beta,$$

where $F_B(1-\alpha; \lambda_1, \lambda_2)$ is the beta distribution function with parameters λ_1 and λ_2 (Wilks, 1941).

Distribution free tolerance interval theory gives r as the smallest positive integer which satisfies the relation

$$F_B(\alpha; B-r+1, r) \geq \beta. \tag{3.29}$$

We will use this result to amend the bootstrap matched filter algorithm of Table 3.7. The only alteration is in Step 5 of the algorithm. The equivalent tolerance interval bootstrap matched filter is shown in Table 3.10.

Example 3.6.3 Performance of tolerance interval matched filter

Consider the same sinusoidal signal as in Examples 3.6.1 and 3.6.2. Suppose that the nominal $\beta = 0.95$ and that the SNR= 5 dB. In Table 3.11, the estimated β using bootstrap detectors with and without tolerance intervals are shown through a Monte Carlo analysis. We used standard Gaussian and Laplace distributed noise in the example.

We have shown that bootstrap detectors, on average, keep the false alarm rate close to the set level when the underlying noise distribution is unknown and/or non-Gaussian. This property of bootstrap detectors is also visible when the noise exhibits non-stationarity. Furthermore, using distribution free tolerance intervals we can specify our confidence that the false alarm rate will not exceed the preset level. The accuracy of tolerance interval bootstrap detection was verified through simulations. We have also found that with real non-Gaussian non-stationary data the tolerance interval based bootstrap detector maintains the set level.

Table 3.10. *The algorithm for the tolerance interval bootstrap matched filter.*

Step 0. *Experiment.* Run the experiment and collect the data x_t, $t = 1, \ldots, n$.

Step 1. *Estimation.* Compute the least square estimator $\hat{\theta}$ of θ, $\hat{\sigma}_{\hat{\theta}}$, and

$$T_n = \left. \frac{\hat{\theta} - \theta}{\hat{\sigma}_{\hat{\theta}}} \right|_{\theta=0}.$$

Step 2. *Resampling.* Compute the residuals

$$\hat{w}_t = x_t - \hat{\theta} s_t, \qquad t = 1, \ldots, n,$$

and after centring, resample the residuals, assumed to be iid, to obtain $\hat{w}_t^*$, $t = 1, \ldots, n$.

Step 3. *Bootstrap test statistic.* Compute new measurements

$$x_t^* = \hat{\theta} s_t + \hat{w}_t^*, \qquad t = 1, \ldots, n,$$

the least squares estimator $\hat{\theta}^*$ based on the resamples x_t^*, $t = 1, \ldots, n$, and

$$T_n^* = \frac{\hat{\theta}^* - \hat{\theta}}{\hat{\sigma}_{\hat{\theta}^*}^*}.$$

Step 4. *Repetition.* Repeat Steps 2 and 3 a large number of times to obtain $T_{n,1}^*, \ldots, T_{n,B}^*$.

Step 5. *Bootstrap test.* Let α be the level of significance, β, be the probability that this level is kept and B be the number of bootstrap resamples. Sort $T_1^*, \ldots, T_B^*$ into increasing order to obtain $T_{(1)}^* \leq \cdots \leq T_{(B)}^*$. Reject H if $T_n \geq T_{n,(r)}^*$, where r is the smallest positive integer which satisfies $F_B(\alpha; B - r + 1, r) \geq \beta$.

3.7 Summary

We have covered the topic of hypothesis testing and signal detection. First, we have reviewed the basics of hypothesis testing and given some examples of classical signal detection. We then have introduced bootstrap methods for signal detection when little is known about the distribution of the interference. The bootstrap matched filter and the CFAR bootstrap matched filter have been introduced and theoretical results in view of their accuracy have also been given. The chapter closed with an interesting application of tolerance intervals which shows how one can incorporate additional probabilistic control of the false alarm rate into the bootstrap matched filter. Examples throughout the chapter show how powerful signal detection with

Table 3.11. *Estimated probability of detection using bootstrap detectors with and without tolerance intervals.*

	β (nominal 0.95) Gaussian	Laplace
Boot. MF	0.5146	0.4003
Boot. CFAR MF	0.4684	0.5549
Tol. Int. Boot. MF	0.9643	0.9398
Tol. Int. Boot. CFAR MF	0.9584	0.9734

the bootstrap can be in the absence of knowledge about the distribution of the interference.

It is worthwhile noting that pivoting is important to ensure high accuracy of bootstrap methods in signal detection. This can be achieved in two ways. One can studentise the statistics in question, which comes at the cost of more computations. Herein, it is important to estimate the variance of the statistic using either the bootstrap or jackknife in the absence of any knowledge of variance. The alternative to pivoting is to variance stabilise the statistics. When a variance stabilising transformation is not available, one can use the bootstrap to achieve this. The examples presented in this chapter show the efficacy of this approach, in view of computations and accuracy.

Alternatively to pivoting, we presented a regression-based method which applies to a linear model. The procedure is more complicated than the pivot method and requires a parametric model so that the bootstrap test can be constructed under the null hypothesis.

The examples with the bootstrap matched filter, bootstrap CFAR matched filter, and the tolerance interval matched filter demonstrate the good performance of the bootstrap in signal detection, with minimal assumptions.

4

Bootstrap model selection

Many engineering applications require parametric modelling. On one hand, there exist statistical models based on real observations of our physical environment (Hahn and Shapiro, 1967); an example of this is statistical modelling of interference (Jakeman and Pusey, 1976; Middleton, 1999). On the other hand, there exist generic models to describe data, such as autoregressive models, commonly used in both signal processing and time series analysis (see Section 2.1.4). In previous chapters, we focused our attention on estimation problems. Given measurements, it is also of importance to decide which model best fits the data. More often we are required to select a model and perform a conditional estimation of the parameters of interest. When we say select a model we mean choose a particular set of parameters in the given model. The conditional estimation refers then to the estimation of those parameters conditioned on the chosen model.

Bootstrap methods based on residuals can be used to select the best model according to a certain prediction criterion. In this chapter, we consider application of bootstrap model selection methods to both linear and nonlinear models. The methods presented are consistent and in most cases they outperform classical techniques of model selection. We also report on how the methods apply to dependent data models such as autoregressive models.

4.1 Preliminaries

Many signal processing problems involve model selection, including parametric spectrum estimation (Dzhaparidze and Yaglom, 1983), system identification (Ljung, 1987), array processing (Böhme, 1991), radar (Rihaczek, 1985) and sonar (Wax and Kailath, 1985). Other model selection problems include application of the maximum *a posteriori* principle (Djurić, 1997), non-linear system modelling (Zoubir *et al.*, 1997), polynomial phase signal modelling

(Zoubir and Iskander, 2000), neural networks (White and Racine, 2001), and biomedical engineering (Iskander *et al.*, 2001). Among signal processing practitioners two approaches for model selection have gained popularity and are widely used (Porat, 1994). These are Akaike's Information Criterion (AIC) (Akaike, 1970, 1974, 1978) and the Bayesian Information Criterion (BIC) (Akaike, 1977; Schwarz, 1978), which is also known as Rissanen's Minimum Description Length (MDL) (Rissanen, 1983). Other model selection methods exist, including the ϕ criterion (Hannan and Quinn, 1979), the C_p method (Mallows, 1973) and the corrected AIC (Hurvich and Tsai, 1989). Relevant publications on the topic include Shibata (1984); Zhao *et al.* (1986); Rao and Wu (1989) and references therein. In some isolated cases, many methods produce marginally better (from a practical point of view) model selection results than those based on Akaike's and the MDL criteria.

Although there exist many model selection procedures, the development of new techniques that outperform the popular ones is still growing and continues to grow. For example, Shi and Tsai (1998) have recently developed a new method based on the generalised Kullback–Leibler information. This is mainly due to the fact that in practice different criteria of statistically equivalent large-sample performance give different answers.

The objective is to introduce methods for model selection based on the bootstrap in a signal processing framework. Besides the good statistical properties of bootstrap selection procedures there are other reasons for the use of the bootstrap for model selection. In previous chapters we have provided evidence of the power of the bootstrap. Let us summarise the advantages of the bootstrap for model selection before we proceed any further:

- The bootstrap does not make assumptions on the distribution of the noise and performs equally well in the non-Gaussian case.

- Usually, model selection is associated with parameter estimation and inference such as variance or mean squared error estimation of parameter estimators and hypothesis testing (e.g., signal detection). As indicated in Chapter 3, inference based on the bootstrap has proved to be asymptotically more accurate than methods based on the Gaussian assumption. Therefore, it is preferable to use the bootstrap for both model selection and subsequent inference applied to the selected model. This does not involve extra cost because the observations generated by the bootstrap for model selection can be used for inference.

- Bootstrap model selection is not limited to independent data models but can be extended to more complicated structures, such as autoregressive models.

In the next sections we present the general theory of model selection with the bootstrap and explain why the methods are attractive for a signal processing practitioner.

4.2 Model selection

Consider a scenario in which we obtain observations from a finite set of random variables $\boldsymbol{Y} = (Y_1, \ldots, Y_n)'$. Let us collate these observations into a vector $\boldsymbol{y} = (y_1, \ldots, y_n)'$. Suppose that we have a choice among q parameter dependent models $\mathcal{M}_1, \ldots, \mathcal{M}_q$ from which to select, based on $\boldsymbol{y}$. The objective of model selection is to choose the model which best explains the data $\boldsymbol{y}$ given a goodness-of-fit measure.

Assume that a model $\mathcal{M}$ is specified by a probability density function $f(\boldsymbol{y}|\boldsymbol{\theta})$ of $\boldsymbol{Y}$ with $\boldsymbol{\theta} = (\theta_1, \ldots, \theta_p)'$ being a p vector-valued parameter to be estimated based on $\boldsymbol{y}$. Provided the probability density function of the data is known, one may use the maximum likelihood approach in which the log-likelihood function $L_n(\boldsymbol{y}|\boldsymbol{\theta}) = \log f(\boldsymbol{y}|\boldsymbol{\theta})$ is utilised to estimate the parameter vector $\boldsymbol{\theta}$. This is achieved by maximising L_n with respect to $\boldsymbol{\theta}$. The maximum of the log-likelihood function is given by

$$L_0(\boldsymbol{y}) = L_n(\boldsymbol{y}|\hat{\boldsymbol{\theta}}) = \sup_{\boldsymbol{\theta}} L_n(\boldsymbol{y}|\boldsymbol{\theta})$$

For a given set of observations this value depends only on the model $\mathcal{M}$. An intuitive solution to the model selection problem may be as follows. Given $\mathcal{M}_1, \ldots, \mathcal{M}_q$, find for each $\mathcal{M}_k$ the corresponding maximum value $L_{0,k}(\boldsymbol{y})$ of the log-likelihood for $k = 1, \ldots, q$. It is known that this approach fails because it tends to pick the model with the largest number of parameters (Rissanen, 1989; Porat, 1994). A solution with the largest number of parameters is impracticable because it contradicts the parsimony principle.† This problem can be overcome by modifying the log-likelihood function in such a way that "parsimonious" models are favoured while "generous" models are penalised. Criteria based on this principle include Akaike's and the MDL criteria. The statistical properties of these methods under the assumption of Gaussian errors have been well studied, see for example Shibata (1984).

Let us investigate bootstrap alternatives to the classical model selection procedures. With few assumptions, bootstrap model selection procedures are shown to be consistent. Let us first consider the simplest case of a linear model and then we will extend our study to more complicated models.

---

† The principle of parsimony is equivalent to the principle of simplicity, i.e. choose a model with the smallest number of parameters.

4.3 Model selection in linear models

Many practical signal processing modelling problems can be reduced in one way or the other to a linear problem. For example, polynomial curve fitting, Fourier analysis, or even non-linear system identification can be considered as linear (in the parameters) estimation problems (Scharf, 1991; Kay, 1993).

Consider the following linear model

$$Y_t = x_t'\theta + Z_t, \qquad t = 1, \ldots, n, \tag{4.1}$$

where Z_t is a noise sequence of iid random variables of unknown distribution with mean zero and variance σ_Z^2. The iid case is considered here for the sake of simplicity, but the methods presented can be extended to the case where Z_t is a correlated process. A discussion on this will be provided in Section 4.4.

In Equation (4.1), θ is an unknown p vector-valued parameter and x_t is the tth value of the p vector of explanatory variables. The output Y_t is sometimes called the response at x_t, $t = 1, \ldots, n$. The vector x_t can be assumed to be random. This will however affect the resampling schemes discussed below. In our treatment, we omit a random x_t and will only consider the case where x_t is fixed.

The model from (4.1) can be re-written as

$$\mu_t = \mathsf{E}[Y_t|x_t] = x_t'\theta,$$
$$\mathsf{var}[Y_t|x_t] = \sigma_Z^2,$$

$t = 1, \ldots, n$, and in a vector form

$$\boldsymbol{Y} = \boldsymbol{x\theta} + \boldsymbol{Z},$$

where $\boldsymbol{Y} = (Y_1, \ldots, Y_n)'$, the matrix $\boldsymbol{x} = (x_1, \ldots, x_n)'$ is assumed to be of full rank, $\boldsymbol{\theta} = (\theta_1, \ldots, \theta_p)'$ and $\boldsymbol{Z} = (Z_1, \ldots, Z_n)'$.

Let β be a subset of $\{1, \ldots, p\}$, $\boldsymbol{\theta}_\beta$ be a sub-vector of $\boldsymbol{\theta}$ containing the components of $\boldsymbol{\theta}$ indexed by the integers in β, and $\boldsymbol{x}_\beta$ be a matrix containing the columns of $\boldsymbol{x}$ indexed by the integers in β. Then, a model corresponding to β is

$$\boldsymbol{Y} = \boldsymbol{x}_\beta \boldsymbol{\theta}_\beta + \boldsymbol{Z}. \tag{4.2}$$

Let β represent a model from now on. The following example will help to clarify this concept.

Example 4.3.1 The concept of a model

Consider finding an empirical relationship between two measured physical variables. Suppose that we choose a fourth order polynomial to establish this relationship, i.e. we choose

$$y_t = a_0 + a_1 x_t + a_2 x_t^2 + a_3 x_t^3 + a_4 x_t^4 + Z_t, \qquad t = 1, \ldots, n.$$

The number of parameters to be estimated here is $p = 5$ and our full model can be written as $\beta = \{a_0, a_1, a_2, a_3, a_4\}$. The parameters can be estimated using the method of least squares. However, modelling involves selection of a model β from within the set of all available parameters. Thus, one model might be

$$y_t = a_0 + a_2 x_t^2 + a_4 x_t^4 + Z_t, \qquad t = 1, \ldots, n,$$

if the regression does not involve odd indexed parameters. The idea of model selection is to find the model β that best (in some sense) represents the data.

We define the optimal model as the model β_o such that $\boldsymbol{\theta}_{\beta_o}$ contains all non-zero components of $\boldsymbol{\theta}$ only. Because $\boldsymbol{\theta}$ is unknown, $\boldsymbol{\theta}_{\beta_o}$ is also unknown. In Example 4.3.1, this would correspond to the model $\beta_o = \{a_0, a_2, a_4\}$.

The problem of model selection is to estimate β_o based on the observations $y_1, \ldots, y_n$.

4.3.1 Model selection based on prediction

Our treatment is based on an estimator of the mean-squared prediction error. A customary predictor for a future response Y_τ at a given $\boldsymbol{x}_{\beta\tau}$ is $\hat{Y}_{\beta\tau} = \boldsymbol{x}'_{\beta\tau} \hat{\boldsymbol{\theta}}_\beta$, where $\boldsymbol{x}'_{\beta\tau}$ is the τth row of $\boldsymbol{x}_\beta$ and $\hat{\boldsymbol{\theta}}_\beta = (\boldsymbol{x}'_\beta \boldsymbol{x}_\beta)^{-1} \boldsymbol{x}'_\beta \boldsymbol{Y}$ is the least squares estimator based on model (4.2). The mean squared error of predicting Y_τ is given by

$$\Gamma(\beta) = \mathsf{E}[Y_\tau - \hat{Y}_{\beta\tau}]^2,$$

where expectation is over Y_τ and $\hat{Y}_{\beta\tau}$; Y_τ and $Y_1, \ldots, Y_n$ being independent. One can show that

$$\Gamma(\beta) = \sigma^2 + \sigma^2 \boldsymbol{x}'_{\beta\tau} (\boldsymbol{x}'_\beta \boldsymbol{x}_\beta)^{-1} \boldsymbol{x}_{\beta\tau} + \Delta(\beta),$$

where $\Delta(\beta) = [\boldsymbol{x}'_\tau \boldsymbol{\theta} - \boldsymbol{x}'_{\beta\tau} (\boldsymbol{x}'_\beta \boldsymbol{x}_\beta)^{-1} \boldsymbol{x}\boldsymbol{\theta}]^2$. If model β is correct in that $\boldsymbol{\theta}_\beta$ contains all non-zero components of $\boldsymbol{\theta}$ such that for any $\boldsymbol{x}$, $\boldsymbol{x}\boldsymbol{\theta} = \boldsymbol{x}_\beta \boldsymbol{\theta}_\beta$ and $\boldsymbol{x}'_\tau \boldsymbol{\theta} = \boldsymbol{x}'_{\beta\tau} \boldsymbol{\theta}_\beta$, then $\Delta(\beta)$ is identically zero. The optimal model is the smallest set with $\Delta(\beta) = 0$. Therefore, if $\Gamma(\beta)$ were known, then the optimal model would be obtained by minimising $\Gamma(\beta)$ with respect to $\beta \subseteq \{1, \ldots, p\}$.

An obvious estimator of $\Gamma(\beta)$ is given by the so-called residual squared error or *apparent* error (Efron and Tibshirani, 1993),

$$\Gamma_n(\beta) = \frac{1}{n} \sum_{t=1}^{n} \left(Y_t - \boldsymbol{x}_{\beta t}' \hat{\boldsymbol{\theta}}_\beta \right)^2 = \frac{\|\boldsymbol{Y} - \boldsymbol{x}_\beta \hat{\boldsymbol{\theta}}_\beta\|^2}{n}, \tag{4.3}$$

with $\boldsymbol{x}_{\beta t}'$ being the tth row of $\boldsymbol{x}_\beta$ and $\|\boldsymbol{a}\| = \sqrt{\boldsymbol{a}'\boldsymbol{a}}$ for any vector $\boldsymbol{a}$. One can show that the expectation of Equation (4.3) is equivalent to

$$\mathsf{E}[\Gamma_n(\beta)] = \sigma^2 - \frac{\sigma^2 p_\beta}{n} + \Delta_n(\beta),$$

where p_β is the size of $\boldsymbol{\theta}_\beta$, $\Delta_n(\beta) = n^{-1}\boldsymbol{\mu}'(\boldsymbol{I} - \boldsymbol{h}_\beta)\boldsymbol{\mu}$, with $\boldsymbol{\mu} = \mathsf{E}[\boldsymbol{Y}] = (\mu_1, \ldots, \mu_n)'$, $\boldsymbol{I}$ is a $p \times p$ identity matrix and $\boldsymbol{h}_\beta = \boldsymbol{x}_\beta(\boldsymbol{x}_\beta'\boldsymbol{x}_\beta)^{-1}\boldsymbol{x}_\beta'$ is the projection matrix. Note that if β is a correct model, then $\Delta_n(\beta)$ is identically zero. It is known that the estimator $\Gamma_n(\beta)$ is too optimistic, i.e., it tends to underestimate the true prediction error (Efron and Tibshirani, 1993), because the same data are used for "testing" and "training". In the next section we develop a bootstrap estimator as an alternative to the above.

4.3.2 Bootstrap based model selection

A straightforward application of the bootstrap for estimating $\Gamma(\beta)$ suggests the estimator

$$\tilde{\Gamma}_n(\beta) = \frac{1}{n} \sum_{t=1}^{n} \left(Y_t - \boldsymbol{x}_{\beta t}' \hat{\boldsymbol{\theta}}_\beta^* \right)^2 = \frac{\|\boldsymbol{Y} - \boldsymbol{x}_\beta \hat{\boldsymbol{\theta}}_\beta^*\|^2}{n}, \tag{4.4}$$

where $\hat{\boldsymbol{\theta}}_\beta^*$ is the bootstrap analogue of the least-squares estimator $\hat{\boldsymbol{\theta}}_\beta$, calculated in the same manner as $\hat{\boldsymbol{\theta}}_\beta$, but with $(Y_t^*, \boldsymbol{x}_{\beta t})$ replacing $(Y_t, \boldsymbol{x}_{\beta t})$.

To obtain bootstrap observations $\boldsymbol{y}^* = (y_1^*, \ldots, y_n^*)'$ of the random vector $\boldsymbol{Y}^* = (Y_1^*, \ldots, Y_n^*)'$, we proceed as follows. Let $\hat{\boldsymbol{\theta}}$ be the least-squares estimate of $\boldsymbol{\theta}$ and define the tth residual by $\hat{z}_t = y_t - \boldsymbol{x}_{\alpha t}'\hat{\boldsymbol{\theta}}_\alpha$, $t = 1, \ldots, n$, where $\alpha = \{1, \ldots, p\}$. Bootstrap resamples $\hat{z}_t^*$ can be generated by resampling with replacement from $(\hat{z}_t - \hat{z}.)/\sqrt{1 - p/n}$ and computing $y_t^* = \boldsymbol{x}_{\beta t}'\hat{\boldsymbol{\theta}}_\beta + \hat{z}_t^*$, $t = 1, \ldots, n$, where $\hat{z}. = n^{-1}\sum_{t=1}^{n} \hat{z}_t$. We included the divisor $\sqrt{1 - p/n}$ to cater for the bias of $\hat{\sigma}^2 = 1/n \sum_{t=1}^{n}(\hat{z}_t - \hat{z}.)^2$. A more accurate adjustment can be obtained by dividing the centred residuals by $\sqrt{1 - p_n/n}$, where $p_n = p + 1 - n^{-1}\sum_{i=1}^{n}\sum_{j=1}^{n} \boldsymbol{x}_i'(\boldsymbol{x}'\boldsymbol{x})^{-1}\boldsymbol{x}_j$, see, for example, Shao and Tu (1995) for more details.

The estimator (4.4) is too variable because it is based on one single bootstrap sample. An improvement can be achieved by considering an average of $\tilde{\Gamma}_n(\beta)$, given by $\bar{\Gamma}_n(\beta) = \mathsf{E}_*[\tilde{\Gamma}_n(\beta)]$, i.e.,

$$\bar{\Gamma}_n(\beta) = \frac{1}{n}\sum_{t=1}^{n}\mathsf{E}_*\left[Y_t - \boldsymbol{x}'_{\beta t}\hat{\boldsymbol{\theta}}^*_\beta\right]^2 = \mathsf{E}_*\frac{\|\boldsymbol{Y} - \boldsymbol{x}_\beta\hat{\boldsymbol{\theta}}^*_\beta\|^2}{n}. \tag{4.5}$$

In practice, $\bar{\Gamma}_n(\beta)$ also tends to underestimate $\Gamma(\beta)$, (Efron and Tibshirani, 1993). However, a more refined bootstrap estimator of $\Gamma(\beta)$ can be found by first estimating the bias in $\Gamma_n(\beta)$ as an estimator of $\Gamma(\beta)$, and then correct $\Gamma_n(\beta)$ by subtracting the estimated bias. The average difference between the true prediction error and its estimate over data sets $\boldsymbol{x}$, called the average optimism (Efron and Tibshirani, 1993), can be estimated as

$$\hat{e}_n(\beta) = \mathsf{E}_*\left[\frac{\|\boldsymbol{Y} - \boldsymbol{x}_\beta\hat{\boldsymbol{\theta}}^*_\beta\|^2}{n} - \frac{\|\boldsymbol{Y}^* - \boldsymbol{x}_\beta\hat{\boldsymbol{\theta}}^*_\beta\|^2}{n}\right].$$

It can be easily shown that the above expression is equivalent to

$$\hat{e}_n(\beta) = \frac{2\hat{\sigma}^2 p_\beta}{n}.$$

The final bootstrap estimator of $\Gamma(\beta)$ is then given by

$$\hat{\Gamma}_n(\beta) = \frac{\|\boldsymbol{Y} - \boldsymbol{x}_\beta\hat{\boldsymbol{\theta}}_\beta\|^2}{n} + \hat{e}_n(\beta). \tag{4.6}$$

This estimator is superior to $\bar{\Gamma}_n(\beta)$ because it uses the observed $(\boldsymbol{x}'_t, y_t)$, $t = 1,\ldots,n$; averaging only enters into the correction term $\hat{e}_n(\beta)$, while $\bar{\Gamma}_n(\beta)$ averages over data sets drawn from the empirical distribution. Although the estimator $\hat{\Gamma}_n(\beta)$ is better than $\bar{\Gamma}_n(\beta)$, the model selection procedure based on $\hat{\Gamma}_n(\beta)$ is inconsistent and too conservative (Shao, 1996). In the next section we present a consistent bootstrap method.

4.3.3 A consistent bootstrap method

Evaluation of Equation (4.6) leads to

$$\hat{\Gamma}_n(\beta) = \frac{\|\boldsymbol{Z}\|^2}{n} + \frac{\|(\boldsymbol{I} - \boldsymbol{h}_\beta)\boldsymbol{\mu}\|^2}{n} \\ - \frac{\|\boldsymbol{h}_\beta\boldsymbol{Z}\|^2}{n} + \frac{2\boldsymbol{Z}'(\boldsymbol{I} - \boldsymbol{h}_\beta)\boldsymbol{\mu}}{n} + \hat{e}_n(\beta),$$

where $\boldsymbol{Z} = (Z_1,\ldots,Z_n)'$. Under some mild regularity conditions, see Shao (1996), for example,

$$\hat{\Gamma}_n(\beta) = \mathsf{E}[\Gamma_n(\beta)] + o_p(1) \tag{4.7}$$

and

$$\hat{\Gamma}_n(\beta) = \frac{\|\boldsymbol{Z}\|^2}{n} + \frac{2\sigma^2 p_\beta}{n} - \frac{\|\boldsymbol{h}_\beta \boldsymbol{Z}\|^2}{n} + o_p(n^{-1})$$

for an incorrect, i.e., $\boldsymbol{\mu} \neq \boldsymbol{x\theta}$, and a correct model, respectively. Herein, the notation $Y_n = o_p(X_n)$ means that $Y_n/X_n = o_p(1)$, i.e., Y_n/X_n converges to zero in probability. The above result indicates that the model selection procedure based on minimising $\hat{\Gamma}_n(\beta)$ over β is inconsistent in that $\lim_{n \to \infty} \mathsf{P}\{\hat{\beta} = \beta_0\} < 1$, unless $\beta = \{1, \ldots, p\}$ is the only correct model. A consistent model selection procedure is obtained if we replace $\hat{e}_n(\beta)$ by $\hat{e}_m(\beta)$ where m is chosen such that, with $h_{\beta t} = \boldsymbol{x}'_{\beta t}(\boldsymbol{x}'_\beta \boldsymbol{x}_\beta)^{-1}\boldsymbol{x}_{\beta t}$,

$$\frac{m}{n} \to 0 \quad \text{and} \quad \frac{n}{m} \max_{t \leq n} h_{\beta t} \to 0$$

for all β in the class of models to be selected. Then,

$$\hat{\Gamma}_{n,m}(\beta) = \frac{\|\boldsymbol{Z}\|^2}{n} + \frac{\sigma^2 p_\beta}{m} + o_p(m^{-1})$$

when β is a correct model, otherwise $\hat{\Gamma}_{n,m}(\beta)$ is as in Equation (4.7). These results suggest that we use the estimator

$$\hat{\Gamma}^*_{n,m}(\beta) = \mathsf{E}_* \left[\frac{\|\boldsymbol{Y} - \boldsymbol{x}_\beta \hat{\boldsymbol{\theta}}^*_{\beta,m}\|^2}{n} \right], \tag{4.8}$$

where the estimate $\hat{\boldsymbol{\theta}}^*_{\beta,m}$ is obtained from $y^*_t = \boldsymbol{x}'_{\beta t}\hat{\boldsymbol{\theta}}_\beta + \hat{z}^*_t$, $t = 1, \ldots, n$, where $\hat{z}^*_t$ denotes the bootstrap resample from $\sqrt{n/m}(\hat{z}_t - \hat{z}.)/\sqrt{1 - p/n}$. We call the latter *detrending* of the original samples.

To evaluate the ideal expression in (4.8), we use Monte Carlo approximations, in which we repeat the resampling stage B times to obtain $\hat{\boldsymbol{\theta}}^{*(i)}_{\beta,m}$ and $\hat{\Gamma}^{*(i)}_{n,m}(\beta)$, and average $\hat{\Gamma}^{*(i)}_{n,m}(\beta)$ over $i = 1, \ldots, B$. The final estimator is then

$$\bar{\Gamma}^*_{n,m}(\beta) = \sum_{i=1}^{B} \frac{\hat{\Gamma}^{*(i)}_{n,m}(\beta)}{B} = \sum_{i=1}^{B} \frac{\|\boldsymbol{Y} - \boldsymbol{x}_\beta \hat{\boldsymbol{\theta}}^{*(i)}_{\beta,m}\|^2}{Bn}, \tag{4.9}$$

The bootstrap procedure for selecting models in linear regressions is summarised in Table 4.1. The computational expense of the procedure is roughly B times larger than that of Akaike's or the MDL criteria, a negligible increase with today's computational power.

Another way to achieve consistency of the bootstrap method is to choose $m < n$ and calculate $\hat{\boldsymbol{\theta}}^*_{\beta,m}$ based only on m bootstrap resamples $y^*_1, \ldots, y^*_m$ generated from the empirical distribution putting mass $1/n$ on y_t, $t = 1, \ldots, n$. In practice it is hard to predict which method will do better.

Table 4.1. *Procedure for model selection in linear models.*

Step 1. Based on $y_1, \ldots, y_n$, compute the least-squares estimate $\hat{\boldsymbol{\theta}}_\alpha$ and

$$\hat{z}_t = y_t - \boldsymbol{x}'_{\alpha t}\hat{\boldsymbol{\theta}}_\alpha, \quad t = 1, \ldots, n,$$

where $\alpha = \{1, \ldots, p\}$.

Step 2. Resample with replacement from $\sqrt{n/m}(\hat{z}_t - \hat{z}.)/\sqrt{1 - p/n}$ to obtain $\hat{z}_t^*$, where $\frac{m}{n} \to 0$ and $\frac{n}{m} \max_{t \leq n} h_{\beta t} \to 0$ for all β in the class of models to be selected.

Step 3. Compute

$$y_t^* = \boldsymbol{x}'_{\beta t}\hat{\boldsymbol{\theta}}_\beta + \hat{z}_t^*, \quad t = 1, \ldots, n$$

and the least-squares estimate $\hat{\boldsymbol{\theta}}^*_{\beta,m}$ from $(y_t^*, \boldsymbol{x}_{\beta t})$.

Step 4. Repeat Steps 2–3 to obtain $\hat{\boldsymbol{\theta}}^{*(i)}_{\beta,m}$ and

$$\hat{\Gamma}^{*(i)}_{n,m}(\beta) = \frac{\|\boldsymbol{y} - \boldsymbol{x}_\beta\hat{\boldsymbol{\theta}}^{*(i)}_{\beta,m}\|^2}{n}, \quad i = 1, \ldots, B.$$

Step 5. Average $\hat{\Gamma}^{*(i)}_{n,m}(\beta)$ over $i = 1, \ldots, B$ to obtain $\bar{\Gamma}^*_{n,m}$ and minimise over β to obtain $\hat{\beta}_0$.

For small samples, we recommend scaling the residuals. For a sufficiently large sample, we have found that both schemes lead to high power. An optimal m may depend on model parameters and thus may be difficult or even impossible to determine. One guideline for choosing m is that p/m should be reasonably small. For more details on this issue and the choice of scale see the work of Shao (1996).

Let us consider two simple examples of bootstrap model selection in linear models.

Example 4.3.2 Trend estimation

Consider estimating the model for a trend in a stationary iid process of unknown distribution. Let

$$Y_t = \boldsymbol{x}'_t\boldsymbol{\theta} + Z_t, \quad t = 1, \ldots, n,$$

where $\boldsymbol{x}_t = (1, t, \ldots, t^p)'$, $t = 1, \ldots, n$, $\boldsymbol{\theta}$ is the vector of polynomial coefficients chosen to be $\boldsymbol{\theta} = (0, 0, 0.035, -0.0005)'$ and $n = 64$. We simulate Y_t by adding Gaussian and t_3-distributed noise of variance of 1 and 3, respec-

Table 4.2. *Empirical probabilities (in percent) of selecting models for a trend with $\boldsymbol{\theta} = (0, 0, 0.035, -0.0005)'$, embedded in Gaussian and t_3 distributed noise, $n = 64$, $m = 2$ (see Example 4.3.2).*

| Model β | $\mathcal{N}(0,1)$ | | | t_3 | | |
	$\bar{\Gamma}^*$	AIC	MDL	$\bar{\Gamma}^*$	AIC	MDL
$(0, 0, b_2, b_3)$	100	91	98	99	89	98
$(0, b_1, b_2, b_3)$	0	5	1	1	5	1
$(b_0, 0, b_2, b_3)$	0	3	1	0	3	1
(b_0, b_1, b_2, b_3)	0	2	0	0	3	0

tively. A MATLAB procedure that can be used to evaluate the probability of selecting the correct model order is given in Section A1.8. In the procedure we have also included calculation of Akaike's and the MDL criteria for comparison purposes.

The bootstrap procedure uses $B = 100$ and $m = 2$. The minimiser of $\bar{\Gamma}_{n,m}^*(\beta)$ is selected as the optimal model. Table 4.2 shows the empirical probabilities based on 1000 independent Monte Carlo runs. Note that only those models are listed which returned at least one hit. Clearly, in this example the bootstrap outperforms Akaike's and the MDL criteria.

Example 4.3.3 Harmonics in additive noise

We give an example where we estimate the model for a superposition of harmonics in a stationary iid process of unknown distribution. The model has been discussed in the context of estimation by Kay (1993, Example 4.2). Let

$$Y_t = \sum_{k=1}^{M} a_k \cos \frac{2\pi kt}{n} + Z_t, \quad t = 1, \ldots, n.$$

Collect the parameters $a_1, a_2, \ldots, a_M$ into the vector $\boldsymbol{\theta}$ and with

$$\boldsymbol{x} = \begin{pmatrix} \cos(\frac{2\pi}{n}) & \cdots & \cos(\frac{2\pi M}{n}) \\ \cos(\frac{4\pi}{n}) & \cdots & \cos(\frac{4\pi M}{n}) \\ \vdots & \vdots & \vdots \\ 1 & \cdots & 1 \end{pmatrix},$$

write the linear regression

$$\boldsymbol{Y} = \boldsymbol{x}\boldsymbol{\theta} + \boldsymbol{Z}.$$

$\boldsymbol{x}$ is an $n \times M$ matrix with entries $\cos(\frac{2\pi mt}{n})$ for $m = 1, \dots, M$ and $t = 1, \dots, n$.

Table 4.3. *Empirical probabilities (in percent) on selecting models for harmonics with $\boldsymbol{\theta} = (0.5, 0, 0.3, 0)'$, embedded in Gaussian noise for $n = 64$ and signal-to-noise power ratio (SNR) of 5 and 20 dB for two m values. Results obtained with AIC and the MDL criterion are also displayed. Models not selected by any of the methods are not shown (see Example 4.3.3).*

SNR	20 dB ($m = 20$)			5 dB ($m = 30$)		
Model β	$\bar{\Gamma}^*_{n,m}$	AIC	BIC	$\bar{\Gamma}^*_{n,m}$	AIC	BIC
$(a_1, 0, 0, 0)$	0	0	0	6	0	11
$(a_1, 0, 0, a_4)$	0	0	0	1	1	0
$(a_1, 0, a_3, 0)$	100	83	95	86	77	82
$(a_1, 0, a_3, a_4)$	0	8	4	3	12	3
$(a_1, a_2, 0, 0)$	0	0	0	1	0	0
$(a_1, a_2, a_3, 0)$	0	9	1	3	10	4

We choose $a_1 = 0.5, a_3 = 0.3$ and $a_k = 0$ otherwise. We simulate Y_t by adding Gaussian and t_3-distributed noise of variance such that the signal-to-noise power ratio varies between 0 and 20 dB. A MATLAB procedure for evaluating the probability of selecting the correct model is given in Section A1.8.

The results obtained for both noise types are similar. The bootstrap procedure is performed using $M = 4$, $B = 100$ and various m values. The minimiser of $\bar{\Gamma}^*_{n,m}(\beta)$ is selected as the optimal model. Table 4.3 shows the empirical probabilities, based on 1000 independent Monte Carlo runs, on selecting some models. Clearly, in this example the bootstrap outperforms Akaike's and the MDL criterion. It also outperforms the ϕ and the corrected Akaike's criterion (AICC) (Hurvich and Tsai, 1989). For example, for the correct model, we obtained 76% and 85% at 20 dB SNR with the ϕ criterion and the AICC, respectively. For 5 dB SNR the empirical probabilities dropped to 72% and 80% respectively.

A more complex application of the method has been presented previously where the optimal model of a polynomial phase signal was estimated (Zoubir and Iskander, 2000). Note that here we consider model selection which differs from model order selection in that the full model rather than only the order is estimated.

4.3.4 Dependent data in linear models

Let us focus our attention on dependent data models. When the additive noise in the linear model is dependent, we may combine the model selection approach with bootstrap resampling schemes outlined in Section 2.1.4.

Recall from the linear model selection procedure of Table 4.1 that we can form residuals

$$\hat{z}_t = y_t - \boldsymbol{x}'_{\alpha t}\hat{\boldsymbol{\theta}}_\alpha, \quad t = 1, \ldots, n.$$

In the iid data case these residuals are simply used in a classical resampling scheme. However, for dependent data, we will use the concept of the block of blocks bootstrap. To perform resampling, we will select a block length M, and form a matrix by dividing $\hat{z}_t$ into $n_b = n/M$ non-overlapping blocks

$$[\hat{z}_{kl}] = \begin{bmatrix} \hat{z}_1 & \hat{z}_{M+1} & \cdots & \hat{z}_{n-M+1} \\ \vdots & \vdots & \vdots & \vdots \\ \hat{z}_M & \hat{z}_{2M} & \cdots & \hat{z}_n \end{bmatrix},$$

$k = 1, \ldots, M$ and $l = 1, \ldots, n_b$. In this way we can assume that the residuals in each of the rows of $[\hat{z}_l]$ are independent. This allows us to use a classical bootstrap resampling scheme in which we resample the columns of $[\hat{z}_{kl}]$. Then, we rearrange the matrix $[\tilde{z}^*_{kl}]$ (after scaling) into a long vector to form $\tilde{z}^*_t$, $t = 1, \ldots, n$, and continue with the algorithm as indicated in Table 4.1.

Note that such a scheme allows us to individually scale portions of data. In other words, the scale parameter m may be different for each of the columns of matrix $[\hat{z}_{kl}]$. One such problem has been encountered in modelling corneal elevations from the videokeratoscopic data where the variance of the noise increases linearly with the radial distance from the centre of the cornea (Iskander *et al.*, 2004).

4.4 Model selection in nonlinear models

4.4.1 Data model

The principles discussed in the previous section are easily extendible to nonlinear models. We define a nonlinear model by

$$Y_t = g(\boldsymbol{x}_t, \boldsymbol{\theta}) + Z_t, \qquad t = 1, \ldots, n, \tag{4.10}$$

where Z_t is a noise sequence of iid random variables of unknown distribution with mean zero and variance σ_Z^2. The model in (4.10) can also be written as

$$\mu_t = \mathsf{E}[Y_t|\boldsymbol{x}_t] = g(\boldsymbol{x}_t, \boldsymbol{\theta}), \qquad \mathsf{var}[Y_t|\boldsymbol{x}_t] = \sigma_Z^2,$$

for $t = 1, \ldots, n$. Herein, g is a known function. Let $\mathcal{B}$ be a collection of subsets of $\{1, \ldots, p\}$, and let $g_{\beta t}(\boldsymbol{\theta}_\beta) = g_\beta(\boldsymbol{x}_{\beta t}, \boldsymbol{\theta}_\beta)$, where $\beta \in \mathcal{B}$ and g_β is the restriction of the function g to the admissible set of $(\boldsymbol{x}_{\beta t}, \boldsymbol{\theta}_\beta)$. Let $\tilde{\mathcal{B}}$ be the admissible set for $\boldsymbol{\theta}$.

4.4.2 Use of bootstrap in model selection

A consistent bootstrap procedure for selecting β is based on minimising an estimator of the prediction error $\Gamma(\beta)$ with respect to β. The estimator is given by

$$\bar{\Gamma}^*_{n,m}(\beta) = \mathsf{E}_* \left[\sum_{t=1}^{n} \frac{\left(Y_t - g_{\beta t}(\hat{\boldsymbol{\theta}}^*_{\beta,m}) \right)^2}{n} \right],$$

where

$$\hat{\boldsymbol{\theta}}^*_{\beta,m} = \hat{\boldsymbol{\theta}}_\beta + \boldsymbol{m}_\beta(\hat{\boldsymbol{\theta}}_\beta)^{-1} \sum_{t=1}^{n} \hat{z}^*_t \dot{\boldsymbol{g}}_{\beta t}(\hat{\boldsymbol{\theta}}_\beta),$$

$\dot{\boldsymbol{g}}(\boldsymbol{\gamma}) = \partial g(\boldsymbol{\gamma})/\partial \boldsymbol{\gamma}$, $\boldsymbol{m}_\beta(\boldsymbol{\gamma}) = \sum_{t=1}^{n} \dot{\boldsymbol{g}}_{\beta t}(\boldsymbol{\gamma}) \dot{\boldsymbol{g}}_{\beta t}(\boldsymbol{\gamma})'$ and $\hat{\boldsymbol{\theta}}_\beta$ being the solution of

$$\sum_{t=1}^{n} \left(y_t - g_{\beta t}(\boldsymbol{\gamma}) \right) \dot{\boldsymbol{g}}_{\beta t}(\boldsymbol{\gamma}) = 0$$

for all $\boldsymbol{\gamma} \in \tilde{\mathcal{B}}$. A detailed procedure is given in Table 4.4.

The proof for consistency of this procedure requires more regularity conditions than the one in Section 4.3. Specifically, conditions for the asymptotic Gaussianity of $\hat{\boldsymbol{\theta}}_\beta$ and its bootstrap analogue are needed (Shao, 1996). For more details, see Shao and Tu (1995). The performance of this method is highlighted in an example.

Example 4.4.1 Oscillations in noise

Consider the model

$$Y_t = \cos(2\pi(a_1 + a_2 t + a_3 t^2)) + Z_t, \quad t = 1, \ldots, n.$$

In this case $\boldsymbol{\theta} = (a_1, a_2, a_3)'$ and $\mathcal{B} = \{\beta_k, k = 1, 2, 3, 4\}$. Then, for example, $g_{\beta \cdot t}(\boldsymbol{\theta}_{\beta \cdot}) = \cos(2\pi a_1)$ $(a_2, a_3 = 0)$, $g_{\beta \cdot t}(\boldsymbol{\theta}_{\beta \cdot}) = \cos(2\pi(a_1 + a_3 t^2))$ $(a_2 = 0)$, $g_{\beta \cdot t}(\boldsymbol{\theta}_{\beta \cdot}) = \cos(2\pi(a_1 + a_2 t))$ $(a_3 = 0)$ and $g_{\beta \cdot t}(\boldsymbol{\theta}_{\beta \cdot}) = \cos(2\pi(a_1 + a_2 t + a_3 t^2))$ $(a_2, a_3 \neq 0)$. We run simulations at 10 and 20 dB SNR with $n = 128$ and $m = 120$. The coefficients were selected to be $a_1 = 0.5$, $a_2 = 0$ and $a_3 = 0.00015$. The empirical probabilities (based on 100 simulations) are

Table 4.4. *Bootstrap procedure for model selection in nonlinear regression.*

Step 1. With y_t, $t = 1, \ldots, n$, find $\hat{\boldsymbol{\theta}}_\alpha$, the solution of

$$\sum_{t=1}^{n} (y_t - g_{\alpha t}(\boldsymbol{\gamma}))\, \dot{\boldsymbol{g}}_{\alpha t}(\boldsymbol{\gamma}) = 0,$$

for all $\boldsymbol{\gamma} \in \tilde{\mathcal{B}}$ with $\alpha = \{1, \ldots, p\}$.

Step 2. Compute the residuals

$$\hat{z}_t = y_t - g_{\alpha t}(\hat{\boldsymbol{\theta}}_\alpha), \quad t = 1, \ldots, n.$$

Step 3. Get $\hat{z}_t^*$, $t = 1, \ldots, n$, iid samples from the empirical distribution putting mass n^{-1} on each

$$\sqrt{n/m}(\hat{z}_t - \hat{z}.)/\sqrt{1 - p/n}, \quad t = 1, \ldots, n.$$

Step 4. With $\dot{\boldsymbol{g}}(\boldsymbol{\gamma}) = \frac{\partial g(\boldsymbol{\gamma})}{\partial \boldsymbol{\gamma}}$ and $\boldsymbol{m}_\beta(\boldsymbol{\gamma}) = \sum_{t=1}^{n} \dot{\boldsymbol{g}}_{\beta t}(\boldsymbol{\gamma})\dot{\boldsymbol{g}}_{\beta t}(\boldsymbol{\gamma})'$, compute

$$\hat{\boldsymbol{\theta}}_{\beta,m}^* = \hat{\boldsymbol{\theta}}_\beta + \boldsymbol{m}_\beta(\hat{\boldsymbol{\theta}}_\beta)^{-1} \sum_{t=1}^{n} \hat{z}_t^* \dot{\boldsymbol{g}}_{\beta t}(\hat{\boldsymbol{\theta}}_\beta).$$

Step 5. Repeat Steps 3–4 a large number of times to obtain $\hat{\boldsymbol{\theta}}_{\beta,m}^{*(i)}$, $i = 1, \ldots, B$.

Step 6. Compute

$$\bar{\Gamma}_{n,m}^*(\beta) = B^{-1} \sum_{i=1}^{B} \sum_{t=1}^{n} \frac{\left(y_t - g_{\beta t}(\hat{\boldsymbol{\theta}}_{\beta,m}^{*(i)})\right)^2}{n}.$$

Step 7. Minimise $\bar{\Gamma}_{n,m}^*(\beta)$ over β to find $\hat{\beta}$.

---

given in Table 4.5. Similar results were obtained under different conditions and parameter settings. Here again, the bootstrap demonstrates its power for model selection in nonlinear models. Note also that unless the problem can be linearised, nonlinear least squares procedures are used. In such cases, the probability of selecting the correct model order also depends on the initial estimates.

In Section 4.3 we discussed correlated noise in a linear model and proposed the use of the block of blocks bootstrap. Alternatively, we would model the coloured noise sequence as an autoregressive process, for example. Then, the residuals of the autoregressive process could be used for resampling. Resampling autoregressive processes for model selection is discussed below.

Table 4.5. *Empirical probabilities (in percent) of selecting the true model from Example 4.4.1, where β_2 is the true model.*

SNR	β_1	β_2	β_3	β_4
20 dB	0	**100**	0	0
10 dB	0	**98**	0	2

4.5 Order selection in autoregressions

The methods discussed above can be generalised to linear random processes. Here, we consider model order selection of an autoregressive process, where

$$Y_t = \theta_1 Y_{t-1} + \theta_2 Y_{t-2} + \cdots + \theta_p Y_{t-p} + Z_t, \quad t = 1, \ldots, n,$$

with p being the order, θ_k, $k = 1, \ldots, p$, are unknown parameters and Z_t are zero-mean iid random variables with variance σ^2. Let

$$(y_{1-p}, \ldots, y_{-1}, y_0, y_1, \ldots, y_n)'$$

be observations of Y_t and collect the parameters into a vector $\boldsymbol{\theta}$ whose least squares estimate is

$$\hat{\boldsymbol{\theta}} = \boldsymbol{r}_n^{-1} \left(\sum_{t=1}^n y_{t-1} y_t, \ldots, \sum_{t=1}^n y_{t-p} y_t \right),$$

where $\boldsymbol{r}_n^{-1}$ is the $p \times p$ matrix whose (i,j)th element is given by $\sum_{t=1}^n y_{t-i} y_{t-j}$. The residuals are obtained from $\hat{z}_t = y_t - \sum_{k=1}^p \hat{\theta}_k y_{t-k}$, $t = 1, \ldots, n$.

A resampling procedure for estimating the variance of the estimator of the parameter of an AR(1) process has been described in Example 2.1.9. The principle can be used here in a similar fashion to estimate the order of an AR process. We thus select a model β from $\alpha = \{1, \ldots, p\}$, where *each β corresponds to the autoregressive model of order β*, i.e., $Y_t = \theta_1 Y_{t-1} + \theta_2 Y_{t-2} + \cdots + \theta_\beta Y_{t-\beta} + Z_t$, $\beta \in \alpha$. The optimal order is

$$\beta_0 = \max \{k : 1 \leq k \leq p, \theta_k \neq 0\},$$

where p is the largest order. The bootstrap approach is described in Table 4.6.

The procedure described in Table 4.6 is consistent in that $\mathsf{Prob}[\hat{\beta} = \beta_0] \to 1$ as $n \to \infty$, provided m satisfies $m \to \infty$ and $m/n \to 0$ as $n \to \infty$. The proof requires stability of the recursive filter and Cramér's condition, see details in Shao (1996). Note that in Sections 4.3 and 4.4 m was a scalar

Table 4.6. *Procedure for order selection in an AR model.*

Step 1. Resample the residuals $(\hat{z}_t - \hat{z}_\cdot)$ to obtain $\hat{z}_t^*$.

Step 2. Find $\hat{\boldsymbol{\theta}}_{\beta,m}^*$ the least-squares estimate of $\boldsymbol{\theta}_\beta$ under β from

$$y_t^* = \sum_{k=1}^{\beta} \hat{\theta}_k y_{t-k}^* + \hat{z}_t^*, \quad t = 1 - p, \ldots, m,$$

with m replacing n and where the initial bootstrap observations $\{y_{1-2p}^*, \ldots, y_{-p}^*\}$ are chosen to be equal to $\{y_{1-p}, \ldots, y_0\}$.

Step 3. Repeat Steps 1–2 to obtain $\hat{\boldsymbol{\theta}}_{\beta,m}^{*(1)}, \ldots, \hat{\boldsymbol{\theta}}_{\beta,m}^{*(B)}$ and

$$\bar{\Gamma}_{n,m}^*(\beta) = \sum_{i=1}^{B} \sum_{t=1}^{n} \frac{\left(y_t - \sum_{k=1}^{\beta} y_{t-k} \hat{\theta}_{k,m}^{*(i)} \right)^2}{Bn}$$

Step 4. Minimise $\bar{\Gamma}_{n,m}^*(\beta)$ over β to find $\hat{\beta}$.

---

of the residuals while here it determines the size of the data used for the bootstrap estimates.

Example 4.5.1 Order selection in an autoregressive model

In this example, we consider the problem of determining the order of the process described by

$$Y_t = -0.4Y_{t-1} + 0.2Y_{t-2} + Z_t, \quad t \in \mathbb{Z},$$

where Z_t is a standard Gaussian variable. A number of $n = 128$ observations was considered. Results of the procedure described in Table 4.6 as well as a comparison with Akaike's and the MDL criterion are given in Table 4.7.

Table 4.7. *Empirical probabilities (in percent) of selecting the true AR model from Example 4.5.1, $p = 2$, $n = 128$ and $m = 40$.*

Method	$\beta = 1$	$\beta = 2$	$\beta = 3$	$\beta = 4$
Bootstrap	28.0	**65.0**	5.0	2.0
AIC	17.8	**62.4**	12.6	7.2
MDL	43.2	**54.6**	2.1	0.1

Similar results were obtained with different constellations and noise types. In these simulations, the choice of m did not appear to have an effect on the results, as long as it satisfied the condition given above.

4.6 Detection of sources using bootstrap techniques

Detection of the number of sources is a preliminary step in array processing problems. It is a model selection problem and so classical techniques such as Akaike's and the MDL criterion have been applied, though the former is not recommended as it consistently over-estimates the number of sources. Several modifications to the MDL criterion have been developed to improve performance for the practical cases of small sample sizes or low SNR (Fishler and Messer, 1999).

The problem is best described in terms of a statistical model where n snapshots of iid zero mean complex data are received from a p element array,

$$\boldsymbol{x}_t = \boldsymbol{A}\boldsymbol{s}_t + \boldsymbol{v}_t, \qquad t = 1, \ldots, n. \tag{4.11}$$

$\boldsymbol{A}$ is a $p \times q$ array steering matrix, $\boldsymbol{s}_t$ is a q ($q < p$) vector valued white source signal and $\boldsymbol{v}_t$ is noise with covariance $\sigma^2 \boldsymbol{I}$. The source and noise are assumed to be independent, so that the covariance of the received signal, the array covariance, is

$$\boldsymbol{R} = \mathsf{E}\left[\boldsymbol{x}_t \boldsymbol{x}_t^H\right] = \boldsymbol{A}\boldsymbol{R}_s \boldsymbol{A}^H + \sigma^2 \boldsymbol{I} \tag{4.12}$$

where $\boldsymbol{R}_s = \mathsf{E}\left[\boldsymbol{s}_t \boldsymbol{s}_t^H\right]$ is the covariance of the sources. The ordered eigenvalues of $\boldsymbol{R}$ are

$$\lambda_1 \geq \cdots \geq \lambda_q > \lambda_{q+1} = \cdots = \lambda_p = \sigma^2 \tag{4.13}$$

so that the smallest $p - q$ population eigenvalues are equal. These equal eigenvalues belong to the noise subspace of the signal, the remaining eigenvalues are said to belong to the signal subspace. The problem of detecting the number of sources is then one of determining the multiplicity of the smallest eigenvalues.

The problem of testing for equality of eigenvalues was considered by Lawley who worked within a hypothesis testing framework to develop the sphericity test (Lawley, 1956). For source detection it is necessary to test in a sequential manner for equality of subsets of the eigenvalues. A modification of Lawley's sphericity statistic has since been developed to account for the presence of eigenvalues not included in the subset being tested and, in contrast to the sphericity statistic, converges to the correct asymptotic χ^2 density giving higher detection rates (Williams and Johnson, 1990).

For finite samples the ordered sample eigenvalues l_i, $i = 1, \ldots, p$, are estimated from the sample covariance

$$\hat{\boldsymbol{R}} = \frac{1}{n-1} \sum_{t=1}^{n} \boldsymbol{x}_t \boldsymbol{x}_t^H \qquad (4.14)$$

and are distinct with probability one for finite sample sizes (Anderson, 1984),

$$l_1 > \cdots > l_q > l_{q+1} > \cdots > l_p > 0. \qquad (4.15)$$

The sample eigenvalues are biased and mutually correlated. Their finite sample joint distribution is known in the Gaussian case and is represented as a series of zonal polynomials (James, 1960), a form too cumbersome for general use. A mathematically tractable form for their asymptotic joint distribution does exist in the Gaussian case (Anderson, 1963), though it may be unreliable for the small sample sizes considered here. Also, this joint distribution is sensitive to departures from Gaussianity (Waternaux, 1976).

All the aforementioned methods are based on the assumption of Gaussian data, that is Gaussian sources and Gaussian noise. Under non-Gaussianity their behaviour is uncertain, at best one can expect their performance to degrade gracefully, though it is known that the distribution of the sample eigenvalues can be sensitive to departures from Gaussianity (Waternaux, 1976). While a reformulation of AIC, MDL or sphericity tests to deal with non-Gaussian data is possible, the multitude of alternatives makes this approach problematic.

Using the bootstrap to overcome the assumption of Gaussianity and large sample sizes, a hypothesis test for equality of the smallest eigenvalues can be developed. The procedure differs from the sphericity test in that all pairwise comparisons, or differences, of the eigenvalues are considered and then combined in a multiple test procedure.

The null distributions of these statistics are estimated using the bootstrap. This means the finite sample and not the asymptotic distributions are estimated, in contrast to the asymptotically correct distributions used in the sphericity test. This makes only minimal assumptions on the distribution of the signal.

4.6.1 Bootstrap based detection

Whether the source detection scheme is based on information theoretic criteria or hypothesis tests, the common assumption of Gaussian data generally leads to some variant of the statistic

$$
\frac{\left(\prod_{i=k}^{p} l_i\right)^{\frac{1}{p-k+1}}}{\frac{1}{p-k+1}\sum_{i=k}^{p} l_i}, \qquad k = 1, \ldots, p-1, \tag{4.16}
$$

which is the ratio of the geometric mean to the arithmetic mean of the smallest $p - k + 1$ sample eigenvalues. For the MDL the term enters as the likelihood function of the data, while for the sphericity test it comprises the test statistic, up to a multiplying factor. When viewed in this sense, the MDL can be considered a type of sphericity test where the level is adaptive. For non-Gaussian data this statistic still has relevance, but the performance of a test based on it is uncertain.

Equation (4.13) suggests q be estimated by determining the multiplicity of the smallest ordered sample eigenvalues. This can be accomplished by considering the following set of hypothesis tests,

$$
\begin{array}{llll}
\mathsf{H}_0 & : & \lambda_1 & = \cdots = \lambda_p \\
& \vdots & & \vdots \\
\mathsf{H}_k & : & \lambda_{k+1} & = \cdots = \lambda_p \\
& \vdots & & \vdots \\
\mathsf{H}_{p-2} & : & \lambda_{p-1} & = \lambda_p
\end{array} \tag{4.17}
$$

with corresponding alternatives K_k : not H_k, $k = 0, \ldots, p-2$. Acceptance of H_k leads to the estimate $\hat{q} = k$. A practical procedure to estimate q starts with testing H_0 and proceeds to the next hypothesis test only on rejection of the hypothesis currently being tested. Upon acceptance the procedure stops, implying all remaining hypotheses are true. The procedure is outlined in Table 4.8.

If one takes the case of Gaussian signals and makes simplifications to obtain an asymptotically correct closed form expression for the distribution of a test statistic based on (4.16), the sphericity test is obtained. Similarly, by following an information theoretic approach one arrives at the MDL.

At this point the sphericity test assumes Gaussianity, here this assumption is not made. Following through with the hypothesis testing approach, con-

Table 4.8. *Hypothesis test for determining the number of sources.*

Step 1. Set $k = 0$.

Step 2. Test H_k.

Step 3. If H_k is accepted then set $\hat{q} = k$ and stop.

Step 4. If H_k is rejected and $k < p - 1$ then set $k \leftarrow k + 1$ and return to Step 2. Otherwise set $\hat{q} = p - 1$ and stop.

---

sider all possible pairwise differences among the sample eigenvalues which lead to the test statistics

$$T_{n,ij} = l_i - l_j, \qquad i = k+1, \ldots, p-1, \quad j = i+1, \ldots, p. \qquad (4.18)$$

These differences will be small when l_i and l_j are both noise eigenvalues, but relatively large if one or both of l_i and l_j are source eigenvalues. Representing the pairwise comparisons in a hypothesis testing framework gives

$$\begin{aligned} \mathsf{H}_{ij} &: \lambda_i = \lambda_j, & & (4.19) \\ \mathsf{K}_{ij} &: \lambda_i \neq \lambda_j, & i = k+1, \ldots, p-1, \quad j = i+1, \ldots, p. \end{aligned}$$

The hypotheses H_k can then be formulated as intersections between the pairwise comparisons,

$$\begin{aligned} \mathsf{H}_k &- \cap_{i,j} \mathsf{H}_{ij}, & & (4.20) \\ \mathsf{K}_k &= \cup_{i,j} \mathsf{K}_{ij}, & i = k+1, \ldots, p-1, \quad j = i+1, \ldots, p. \end{aligned}$$

The pairwise comparisons are carried out using a multiple test procedure to maintain a global level of significance, as discussed next.

4.6.1.1 Multiple hypothesis testing

Multiple hypothesis tests are employed when several individual hypotheses are tested simultaneously and a certain global level of significance must be maintained (Hochberg and Tamhane, 1987). The global level of significance, α, is defined as the probability with which at least one of the hypotheses is rejected, given that all are true,

$$\alpha = \mathsf{Pr}(\text{reject at least one } \mathsf{H}_{ij} \mid \text{all } \mathsf{H}_{ij} \text{ are true}). \qquad (4.21)$$

In the above test for equality of eigenvalues the global hypothesis, H_0, is that all eigenvalues are equal. Thus, when no sources are present, the probability of correctly choosing $\hat{q} = 0$ should be maintained at $1 - \alpha$.

Bonferroni's multiple test procedure is one of the simplest such methods. Each hypothesis, H_{ij}, is tested at a level α/h where $h = p(p-1)/2$ is the number of hypotheses comprising the global hypothesis. Assuming the significance values are independent and uniformly distributed on $[0, 1)$ this method exactly controls the level at α. When the significance values are dependent, Bonferroni's procedure is conservative and has reduced power. Stepwise methods such as Holm's sequentially rejective Bonferroni procedure (SRB) are less conservative while strongly controlling the level. Strong control is a favourable property implying (4.21) is satisfied over all subsets of hypotheses including the global null. For further details on multiple hypothesis tests see Hochberg and Tamhane (1987).

In this problem the significance values are not independent due to correlation between sample eigenvalues and logical implications between hypotheses. For instance, if H_{1p} were true, this would imply all the H_{ij} were true. Hence, the SRB procedure is used exclusively. The procedure as it applies to this problem is as follows. Given significance values P_{ij} for the H_{ij}, order them so that $P_{(1)} \geq P_{(2)} \geq \cdots \geq P_{(h)}$ and label the corresponding hypotheses $\mathsf{H}_{(1)}, \ldots, \mathsf{H}_{(h)}$. If $P_{(h)} \geq \alpha/h$ then all hypotheses are accepted, otherwise $\mathsf{H}_{(h)}$ is rejected and $\mathsf{H}_{(h-1)}$ is tested. If $P_{(h-1)} \geq \alpha/(h-1)$ then the hypotheses $\mathsf{H}_{(1)}, \ldots, \mathsf{H}_{(h-1)}$ are accepted, otherwise $\mathsf{H}_{(h-1)}$ is rejected and $\mathsf{H}_{(h-2)}$ is tested. The procedure continues until either the remaining hypotheses are accepted, or all are rejected. Table 4.9 summarises the method.

Table 4.9. *Holm's sequentially rejective Bonferroni procedure.*

Step 1. Set $i = q$.
Step 2. If $P_{(i)} \geq \alpha/i$, accept all $\mathsf{H}_{(i)}, \ldots, \mathsf{H}_{(1)}$ and stop, otherwise reject $\mathsf{H}_{(i)}$.
Step 3. If $i = 1$ stop, otherwise set $i \leftarrow i - 1$ and return to Step 2.

Once the H_{ij} have been tested, all that remains is to step through Table 4.8. From (4.20) it is evident that H_k is rejected if any of the hypotheses H_{ij}, $i = k+1, \ldots, p$, $j > i$, are rejected. Significance values are determined from estimates of the null distributions, obtained using the bootstrap, as discussed next. A similar approach was considered in Zoubir and Böhme (1995) to find significance values for complicated test statistics.

4.6.2 Null distribution estimation

Evaluation of the significance values needed to carry out the hypothesis tests requires knowledge of the null distributions of the test statistics, the pairwise differences, $T_{n,ij} = l_i - l_j$. The bootstrap is used as a nonparametric estimator of the null distributions (Efron and Tibshirani, 1993). Using the bootstrap allows us to avoid making assumptions about the distribution of the signals. As mentioned, this advantage is quite important when working with eigenvalues since their distribution is too complex for general use (James, 1960), while asymptotic expansions derived under Gaussianity (Anderson, 1963) may not be valid for the small sample sizes considered. Asymptotic approximations developed for non-Gaussian cases require knowledge of the higher order moments of the data, which are difficult to estimate for small sample sizes (Waternaux, 1976; Fujikoshi, 1980). Next we present the bootstrap method as applied to this problem.

Bootstrapping eigenvalues. Here, the bootstrap is used to estimate the distribution of the test statistics from the sample. The principle is to treat the sample as an empirical estimate of the true distribution and then to resample from this estimate, creating bootstrap data sets. These data sets can then be used to perform inference.

In this case the sample is a set of vectors, the array snapshots, collected into the data matrix

$$\boldsymbol{X} = [\boldsymbol{x}_1, \ldots, \boldsymbol{x}_n]. \tag{4.22}$$

Let the empirical density be given by delta functions located at each sample and weighted by $1/n$. Resampling from this empirical density is equivalent to randomly resampling from the original data set with replacement, giving the bootstrap data

$$\boldsymbol{X}^* = [\boldsymbol{x}_1^*, \ldots, \boldsymbol{x}_n^*]. \tag{4.23}$$

The sample eigenvalues of this data set are

$$l_1^* > \cdots > l_p^*. \tag{4.24}$$

From the bootstrapped eigenvalues we can form the required test statistic. Repeating the procedure B times gives an estimate of the distribution of this test statistic which can be used for inference. The bootstrap procedure for eigenvalues is shown in Table 4.10.

For linear statistics, such as the sample mean, the bootstrap is known to perform well and is a consistent estimator of the distribution. For complex nonlinear operations such as eigenvalue estimation these properties may not

Table 4.10. *Bootstrap procedure for resampling eigenvalues.*

Step 1. Define the matrix of array snapshots $X = [x_1, \ldots, x_n]$.

Step 2. Randomly select a snapshot from X with replacement. Repeat n times to form the bootstrap data $X^* = [x_1^*, \ldots, x_n^*]$.

Step 3. Centre X^* by subtracting the sample mean from each row,

$$x_i^* \leftarrow x_i^* - \frac{1}{n}\sum_{j=1}^{n} x_j^*, \qquad i = 1, \ldots, n.$$

Step 4. Estimate the sample correlation matrix, $\hat{R}^*$, of the centred X^*.

Step 5. The resampled eigenvalues, $l_1^*, \ldots, l_p^*$, are estimated from $\hat{R}^*$.

Step 6. Repeat Steps 2 to 5 B times to obtain the bootstrap set of eigenvalues $l_1^*(b), \ldots, l_p^*(b)$, $b = 1, \ldots, B$.

---

apply. In Beran and Srivastava (1985, 1987) the statistical properties of the bootstrapped eigenvalues are considered. It is shown that while the bootstrap converges to the correct asymptotic distributions for distinct eigenvalues, the same is not true of the multiple eigenvalues. Bootstrapping with fewer resamples, m, where $m < n$, $\min(m, n) \to \infty$ and $m/n \to 0$, ensures the bootstrap converges weakly to the asymptotic distribution for multiple eigenvalues with sample size m. The sample sizes considered here are quite small, being on the order of 100. It was found that to fulfil the conditions which ensure weak convergence for the multiple eigenvalues the decrease in resample size would increase the error in eigenvalue estimation to the point that the overall error in the distribution increased.

For large sample sizes subsampling may be used to ensure weak convergence of the multiple eigenvalues to their asymptotic distributions. The conditions under which subsampling is valid encompass a wider range of distributions and more complex statistics than with the bootstrap (Politis *et al.*, 1999). In subsampling the data is resampled $m < n$ times, either with or without replacement. To account for the decrease in sample size the rate of convergence to the asymptotic distribution must be estimated. Usually the rate is of the form $(m/n)^\tau$ where $\tau \approx 0.5$. The use of subsampling for this problem was investigated, though again, the bootstrap was found to provide a sufficiently accurate estimate.

Bootstrap Procedure. Given the bootstrap set of eigenvalues the test statistic of (4.18) is recalculated giving $T^*_{n,ij} = l^*_i - l^*_j$. Repeating this procedure B times gives the set of bootstrapped test statistics, $T^*_{n,ij}(b)$, $b = 1, \ldots, B$. From these bootstrapped statistics the distribution of $T_{n,ij}$ under the null hypothesis is estimated as $T^{\mathsf{H}}_{n,ij}(b) = T^*_{n,ij}(b) - T_{n,ij}$ (Efron and Tibshirani, 1993). Note that the test statistics are not studentised, preliminary investigations having shown the extra computational cost required to be unwarranted. Significance values for the two-sided hypothesis tests are then evaluated as

$$P_{ij} = \frac{1}{B} \sum_{b=1}^{B} I\left(|T_{n,ij}| \leq |T^{\mathsf{H}}_{n,ij}(b)|\right) \tag{4.25}$$

where $I(\cdot)$ is the indicator function.

4.6.3 Bias correction

Distinct sample eigenvalues, or those corresponding to sources, are asymptotically unbiased, while the multiple sample eigenvalues corresponding to the noise only are asymptotically biased. In the small sample case the bias becomes quite significant.

The hypothesis tests upon which the detection scheme rests are based upon detecting a statistically significant difference between the eigenvalues. Bias in the multiple sample eigenvalues falsifies the assumption that noise only sample eigenvalues have equal means. This necessitates some form of bias correction.

Several resampling methods for bias correction exist based on the bootstrap, the jackknife and subsampling (Efron and Tibshirani, 1993; Politis, 1998). Aside from the problems of non-Gaussianity and small sample sizes, the advantage of resampling techniques for bias correction in this case is that they can be applied blindly, with no knowledge of the eigenvalue multiplicity.

In Brcich *et al.* (2002) it was found that the jackknife estimate of bias *Bias*$_{\mathsf{JCK}}$ performed best among the resampling methods. It also performed well compared to a robust estimator of eigenvalue bias, *Bias*$_{\mathsf{LBE}}$, developed from the expected value of the sample eigenvalues as found by Lawley (1956). *Bias*$_{\mathsf{LBE}}$ is defined as

$$\widehat{Bias}_{\mathsf{LBE}}(l_i) = \frac{1}{N} \sum_{j=i+1}^{p} l_j \sum_{k=0}^{K} \left(\frac{l_j}{l_i}\right)^k - \frac{1}{N} \sum_{j=1}^{i-1} l_i \sum_{k=0}^{K} \left(\frac{l_i}{l_j}\right)^k \tag{4.26}$$

for some suitable K. Limiting K guards against a large increase in the variance of the bias corrected eigenvalues when they are not well separated. This must be balanced against an increase in the bias of the estimator for small K. A moderate value of $K = 25$ was used here.

Note that bias correction is applied to both the sample eigenvalues and the bootstrapped eigenvalues. A summary of the entire detection procedure is given in Table 4.11.

Table 4.11. *Bootstrap detection procedure.*

Step 1. Estimate the eigenvalues, $l_1, \ldots, l_p$, from the matrix of array snapshots and apply one of the bias corrections $Bias_{\text{LBE}}$ or $Bias_{\text{JCK}}$.
Step 2. Obtain the B bootstrapped eigenvalue sets as in Table 4.10 and bias correct each set using the same correction procedure as above.
Step 3. Calculate the test statistics $T_{n,ij}$ and the bootstrap estimate of their distributions under the null, $T_{n,ij}^{\text{H}}$.
Step 4. Given the level, carry out the multiple hypothesis test as in Table 4.8.

4.6.4 Simulations

The following simulation compares bootstrap based detection with the MDL (Wax and Kailath, 1985) and the sphericity test (Williams and Johnson, 1990). In this scenario $q = 3$ sources at 10, 30 and 50 degrees and SNRs of -2, 2 and 6 dB respectively impinge on a $p = 4$ element array whose element spacing which was one half the wavelength, the signals were Gaussian. The number of resamples was $B = 200$ and the global level of significance $\alpha = 2\%$.

For small sample sizes of 100 snapshots or less the bootstrap method achieves a consistently higher detection rate than the sphericity test (see Figure 4.1). As already discussed, this is a result of the bootstrap estimating the finite sample distributions in contrast to the asymptotic correctness of the sphericity test. Further results can be found in Brcich *et al.* (2002).

4.7 Summary

Throughout this chapter we have focused our attention on model selection. Linear as well as nonlinear models have been considered. We have examined the case of deterministic explanatory variables only. However, as mentioned earlier, the extension to a random explanatory variable is straightforward

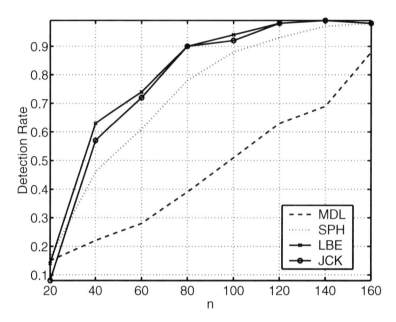

Fig. 4.1. Empirical probability of correctly detecting the number of sources as the sample size is varied. JCK denotes Jackknife bias correction and LBE Lawley's modified correction. Results were averaged over 100 Monte Carlo simulations.

and requires resampling pairs or cases. This type of resampling leads to a design matrix x^* that is not equal to the original x. This is not problematic for large samples. However, for a small n the scheme has to be viewed with caution. A good feature of resampling pairs (even for a deterministic x) is that the scheme is robust because it caters for model uncertainty. More details on this issue can be found in the work of Shao (1996).

We have also emphasised that bootstrap model selection procedures will not be consistent unless the residuals are appropriately scaled. This may cause a problem in those practical applications in which the level of noise changes or is difficult to predict. We have observed in our experimental work that there is a strong link between the signal-to-noise ratio and the optimal level of scaling. For example, if the signal to noise ratio is small, we choose a smaller m to increase the variability of the residuals. In the alternative case, the scaling parameter should be larger so that the model is not overparameterised. Thus, the choice of m is crucial. On the other hand, we have experienced many cases in which the model selection algorithm performs well for a wide range of scaling parameters. In summary, in many cases it is up to the designer to evaluate the scaling parameter through trial and error because no theory for optimal selection of the parameter m exists.

We have also reported bootstrap methods for model order selection in autoregressions. All the methods presented in this chapter show that they outperform classical model selection techniques such as Akaike's Information Criterion and Rissanen's Minimum Description Length. This is mostly evident in small sample cases.

The chapter closes with an application of the bootstrap to the detection of the number of sources impinging on an array. The methods outperform existing techniques, including the sphericity test, but at the cost of additional computation.

5

Real data bootstrap applications

Chapters 3 and 4 dealt with fundamentals of bootstrap based detection and model selection, respectively. In this chapter, we provide some interesting applications of the theory covered in the former chapters to real world problems. We report only on some problems we worked on over the last years. These selected problems had been solved using classical techniques only if we had made strong assumptions which may not be valid. They are also analytically intractable.

The applications include a wide range of signal processing problems. We first report on results for optimal vibration sensor placement on spark ignition engines to detect knock. We show how the bootstrap can be used to estimate distributions of complicated statistics. Then we discuss a passive acoustic emission problem where we estimate confidence intervals for an aircraft's flight parameters. This is followed by the important problem of civilian landmine detection. We suggest an approach to detect buried landmines using a ground penetrating radar. We continue with another radar application concerning noise floor estimation in high frequency over-the-horizon radar. The chapter concludes with the estimation of the optimal model for corneal elevation in the human eye.

5.1 Optimal sensor placement for knock detection

This application illustrates the concepts discussed in Sections 3.2 and 2.2 of hypothesis testing and variance stabilisation, respectively. For a more detailed treatment of this application the reader is referred to Zoubir and Böhme (1995).

5.1.1 Motivation

A means for lowering fuel consumption in spark ignition engines is to increase the compression ratio. This is, however, limited by the occurrence of knock, an extremely fast combustion that generates a knocking or ringing sound. Knock has a damaging effect on the engine, especially when it occurs at high speed. It also reduces efficiency due to heat loss resulting from turbulence in the combustion chamber (König and Böhme, 1996). Once detected, knock can be controlled by adapting the angle of ignition.

A means of detecting knock accurately is to record the in-cylinder pressure via pressure transducers mounted at suitable points inside the cylinder. However, installation of pressure transducers is costly and their use is restricted to reference purposes in engine or fuel development. Alternatively, the use of vibration sensors mounted on the engine block (currently installed in serial productions) for knock detection is economical and easy. However, the performance of knock detection systems suffers from the low signal-to-noise power ratio of structural vibration signals due to noise from mechanical sources such as valves or bearings.

To improve the detection of knock at moderate cost advanced signal processing methods have been used to analyse structural vibrations. Further improvements in detection power are possible if one can find suitable sensor positions for knock detection. Suitably placed vibration sensors improve, independently of the knock detection scheme used, the detectability of knocking cycles and permit safe operation at high efficiency. Below, we illustrate a bootstrap approach to solve the problem of optimal sensor placement.

5.1.2 Data model

The solution lies in finding the optimal distribution of vibration sensors. First, a group of sensors is placed on the engine wall in a heuristic fashion. Next, statistical tests are used to select from among this group of sensors those ones that most accurately model the in-cylinder pressure recorded as a reference.

The real scenario is depicted in Figure 5.1 where a 1.8 l spark ignition engine with four cylinders, a compression ratio of 10 : 1 and 79 kW power was used. The in-cylinder pressure signals of cylinders 1 and 3 (D1 and D3 in Figure 5.1) were recorded together with the output signals of eight acceleration sensors placed at heuristically chosen positions on the engine block and the cylinder head (sensor SK1). The manufacturer's aim is to use only one sensor for knock detection in a four cylinder engine.

The idea is to use a prediction filter as depicted in Figure 5.2. In this

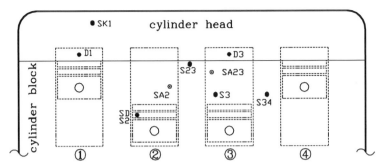

Fig. 5.1. Position of sensors on the engine.

figure, S_t, $t = 0, \pm1, \pm2, \dots$, models the in-cylinder pressure in an arbitrary cylinder (signal of sensor D1 or D3 in Figure 5.1). The r vector-valued process $\boldsymbol{Z}_t$ describes the vibration signals. For this purpose, $\boldsymbol{Z}_t$, $t = 0, \pm1, \pm2, \dots$, is structured into $\boldsymbol{Z}_t = (\boldsymbol{X}'_t, Y_t)'$ where Y_t is the output signal of an arbitrary sensor within the group of r sensors. As described in Figure 5.2, S_t is reconstructed by linearly transforming at once $\boldsymbol{Z}_t$ and $\boldsymbol{X}_t$ (omitting Y_t). On the basis of the resulting error spectral increment $C_{\mathcal{E}_2\mathcal{E}_2}(\omega) - C_{\mathcal{E}_1\mathcal{E}_1}(\omega)$, the irrelevancy of the signal Y_t in predicting S_t at ω is decided. The error increment is proportional to $|R_{SZ}(\omega)|^2 - |R_{SX}(\omega)|^2$ the so-called "coherence gain" contributed by signal Y_t (Zoubir and Böhme, 1995). The lower the coherence gain, the less the contribution of Y_t for approximating S_t from $\boldsymbol{Z}_t = (\boldsymbol{X}'_t, Y_t)'$ by linear time-invariant operations. Thus, the closeness to zero of $|R_{SZ}(\omega)|^2 - |R_{SX}(\omega)|^2$ is a measure of the irrelevancy of the sensor whose output signal is Y_t, which can be formulated as the test of the hypothesis

$$\mathsf{H} : |R_{SZ}(\omega)|^2 - |R_{SX}(\omega)|^2 \le \theta_0(\omega)$$

against the one-sided alternative,

$$\mathsf{K} : |R_{SZ}(\omega)|^2 - |R_{SX}(\omega)|^2 > \theta_0(\omega)$$

where $\theta_0(\omega)$ is a suitably chosen bound close to zero.

Given observations S_t, $\boldsymbol{Z}_t$ for $t = 1, \dots, n$ and L independent combustion cycles, an estimate of the spectral density matrix $\boldsymbol{C}_{VV}(\omega)$ of the $r+1$ vector valued process $\boldsymbol{V}_t = (\boldsymbol{Z}'_t, S_t)'$ is given by

$$\hat{\boldsymbol{C}}_{VV}(\omega) = \frac{1}{L} \sum_{i=1}^{L} \frac{1}{2m+1} \sum_{s=-m}^{m} \boldsymbol{I}_{VV} \left(\frac{2\pi(s(n)+s)}{n}, i \right),$$

where $s(n)$ is an integer with $2\pi s(n)/n$ close to ω, m is a positive integer,

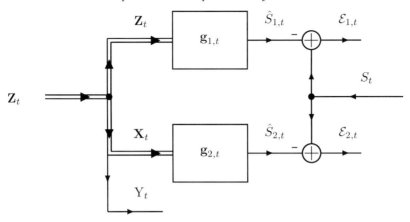

Fig. 5.2. Prediction of the reference signal S_t by linear time-invariant filtering of $\mathbf{Z}_t$ or $\mathbf{X}_t$.

and $\mathbf{I}_{VV}(\omega)$ is the matrix of periodograms. Based on $\hat{\mathbf{C}}_{VV}(\omega)$

$$|\hat{R}_{SZ}(\omega)|^2 = \frac{\hat{\mathbf{C}}_{SZ}(\omega)\hat{\mathbf{C}}_{ZZ}(\omega)^{-1}\hat{\mathbf{C}}_{ZS}(\omega)}{\hat{C}_{SS}(\omega)}$$

and

$$|\hat{R}_{SX}(\omega)|^2 = \frac{\hat{\mathbf{C}}_{SX}(\omega)\hat{\mathbf{C}}_{XX}(\omega)^{-1}\hat{\mathbf{C}}_{XS}(\omega)}{\hat{C}_{SS}(\omega)},$$

are computed.

The density function of $|\hat{R}_{SZ}(\omega)|^2$ was given by Goodman (1963), where $\hat{\mathbf{C}}_{VV}(\omega)$ was assumed to follow a Wishart distribution (see also the work by Khatri (1965)). This holds asymptotically as $n \to \infty$ (Brillinger, 1981, 1983) or for any fixed n given $\mathbf{V}_t$ is Gaussian.

One encounters two problems with coherences. First, the applicability of the asymptotic distribution given in Goodman (1963) and Khatri (1965) becomes questionable when n is small. Second, if one intends to conduct tests for arbitrary nonzero hypothetical values of the multiple coherence, the distribution of the statistic becomes intractable even when assuming n is large. The former problem becomes difficult and the latter even more if one is interested in testing coherence differences such as $|R_{SZ}(\omega)|^2 - |R_{SX}(\omega)|^2$. This is because the sample value $|\hat{R}_{SZ}(\omega)|^2 - |\hat{R}_{SX}(\omega)|^2$ is the difference of two statistically dependent variables ($\mathbf{X}_t$ is a subvector of $\mathbf{Z}_t$). This motivates the use of the bootstrap to conduct accurate tests for the coherence gain.

5.1.3 Bootstrap tests

Let $\theta(\omega) = |R_{SZ}(\omega)|^2 - |R_{SX}(\omega)|^2$ denote the parameter which is esti-
mated by $\hat{\theta}(\omega) = |\hat{R}_{SZ}(\omega)|^2 - |\hat{R}_{SX}(\omega)|^2$, derived from the sample $\mathcal{X}(\omega) =$
$\{d_V(\omega, 1), \ldots, d_V(\omega, L)\}$, where $d_V(\omega, i)$, $i = 1, \ldots, L$, is the normalised
finite Fourier transform of the ith record of the stationary series V_t.

Following the procedures of Table 3.1, and Table 2.3 or (3.12) one would
approximate the distribution of $T_n(\omega) = (\hat{\theta}(\omega) - \theta_0(\omega))/\hat{\sigma}(\omega)$, by the distribution of the bootstrap statistic $T_n^*(\omega) = (\hat{\theta}^*(\omega) - \hat{\theta}(\omega))/\hat{\sigma}^*(\omega)$ and perform
the test of the hypothesis H. In order to ensure that the statistic is pivotal,
we apply a variance stabilising transformation to $\hat{\theta}(\omega)$ prior to the tests, as
illustrated in Table 2.12.

As an example, Figure 5.3 shows the estimated standard deviations of
$B_3 = 1000$ (see Table 2.12) bootstrap estimates of the coherence gain at
the first resonance frequency and 1750 rpm. The solid line results from a
quadratic polynomial fit. It is seen that the standard deviation strongly
depends on the parameter estimates.

The variance stabilising transformation obtained for this experiment is
depicted in Figure 5.4 and was found using $B_1 = 200, B_2 = 25$. In the experiment, a fixed-span running lines smoother with span of 50% to smooth
the $\hat{\sigma}_i^*$ values over the bootstrap estimates $\hat{\theta}_i^*$, $i = 1, \ldots, 1000$, of $\hat{\theta} =$
$|\hat{R}_{SZ}(\omega)|^2 - |\hat{R}_{SX}(\omega)|^2$ was used.

The standard deviations of new $B_3 = 1000$ transformed bootstrap estimates are shown in Figure 5.5. One can see that the obtained parabola in
Figure 5.5 is flatter than the one in Figure 5.3, and so the assumption of a
variance-stable scale is more accurate (Zoubir and Böhme, 1995).

In this application, we have to consider P resonance frequencies and also
$2m + 1$ frequencies in the neighbourhood of each resonance frequency, because the resonance frequencies are not known exactly and because of damping in the knock signals. Thus, the test becomes one of multiple hypotheses
$H_{1,-m}, \ldots, H_{1,m}, \ldots, H_{P,-m}, \ldots, H_{P,m}$ against one-sided alternatives, where
$H_{p,k}$ is given by

$$H_{p,k} : |R_{SZ}(\omega_{p,k})|^2 - |R_{SX}(\omega_{p,k})|^2 \leq \theta_0, \tag{5.1}$$

where the bound θ_0 is suitably chosen in the interval $(0, 1]$ and $\omega_{p,k}$, $k =$
$-m, \ldots, m$, are frequencies in the neighbourhood of the resonance frequen-
cies ω_p with $\omega_{p,0} = \omega_p$, $p = 1, \ldots, P$. For testing the multiple hypotheses,
a bootstrap version of Holm's generalised sequentially rejective Bonferroni
multiple test, as discussed in Section 4.6 is applied (Holm, 1979; Hochberg
and Tamhane, 1987). This procedure consists of evaluating the $\mathcal{P}$-values

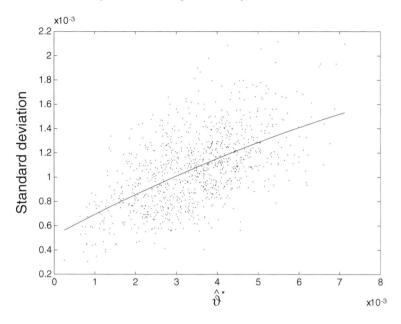

Fig. 5.3. Standard deviation of bootstrap estimates of the coherence gain prior to a variance stabilising transformation.

of the statistic $T_n(\omega)$ at different frequencies and sequentially rejecting single hypotheses. The role of the bootstrap is to approximate the unknown $\mathcal{P}$-values by their bootstrap counterparts (Zoubir, 1994).

5.1.4 The experiment

The experiment was performed at three rotation speeds (1750, 3500 and 5250 rpm) under strong knocking conditions. A large number of combustion cycles of length $n = 128$ data points each were digitised with respect to each cylinder for all three rotation speeds. Then, for each cylinder and each rotation speed $L = 15$ knocking combustion cycles with the highest power were selected.

Four resonance frequencies were found from an estimate of $C_{SS}(\omega)$, obtained by averaging as many periodograms as the number of digitised cycles for each cylinder and each rotation speed. Then, the spectral density matrix $\boldsymbol{C}_{VV}(\omega)$ was estimated by averaging 15 periodograms and smoothing over three frequency bins ($m = 1$). From this estimate, $|\hat{R}_{SZ}(\omega_{p,k})|^2 - |\hat{R}_{SX}(\omega_{p,k})|^2$, $p = 1, \ldots, P$, $k = -m, \ldots, m$, was calculated for all three rotation speeds and two cylinders.

After variance stabilising the sample coherence gain at the frequencies of

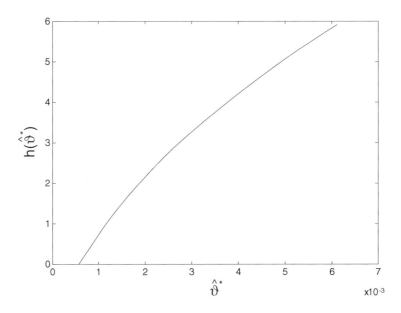

Fig. 5.4. Estimated variance stabilising transformation for the coherence gain.

interest for each cylinder and each rotation speed, bootstrap based multiple tests of the hypotheses (5.1) were performed at a nominal multiple level of significance $\alpha = 5\%$. The irrelevancy of an arbitrary sensor with output signal Y_t was tested by interchanging the components of the vector $\boldsymbol{Z}(t)$ and each time considering the coherence gain $|R_{SZ}(\omega)|^2 - |R_{SX}(\omega)|^2$. Based on the number of rejected hypotheses at each step, sensors were removed one at a time until only one sensor that best approximated the reference was left. Results were encouraging and concurred with the heuristical approach used by the manufacturer (Zoubir and Böhme, 1995).

5.2 Confidence intervals for aircraft parameters

The application reported in this section illustrates the concepts presented in Section 3.2 for confidence interval estimation using the percentile-t method. For more details the reader is referred to work of Zoubir and Boashash (1998).

5.2.1 Introduction

We wish to estimate an aircraft's constant height, velocity, range and acoustic frequency based on a single acoustic recording of the aircraft passing

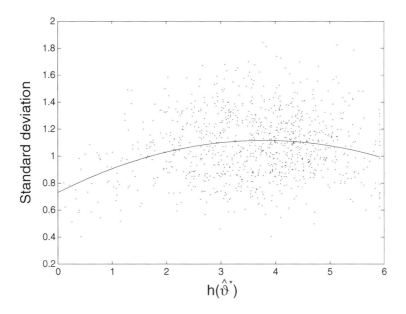

Fig. 5.5. Standard deviation of bootstrap estimates of the coherence gain after a variance stabilising transformation.

overhead. Information about the physical parameters is contained in the phase and time varying Doppler frequency shift, or instantaneous frequency (IF), of the observed acoustic signal. The passive acoustic scenario is depicted in Figure 5.6.

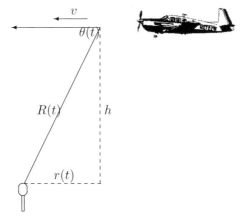

Fig. 5.6. Schematic of the geometric arrangement of the aircraft and observer in terms of the physical parameters of the model. It is assumed that the aircraft is travelling at the constant speed v and at constant height h.

An estimation scheme for the parameters of interest has been previously demonstrated (see, for example, Ferguson (1992); Ferguson and Quinn (1994); Reid *et al.* (1997)), using a model for the IF.

In this problem, we are not only interested in estimating the parameters, but in establishing some statistical confidence for the parameters of interest based on those estimates. Ideally these bounds are found by determining the theoretical distribution of the parameter estimates. However, the theoretical derivation of the parameter distribution is mathematically intractable, particularly when the distribution of the noise is unknown. Alternatively, if multiple realisations of the acoustic signal were available it would be straightforward to empirically determine the distribution of the parameter estimates. In practice this is not possible since only a single realisation of the acoustic signal is available. The alternative is to use the bootstrap as shown below, without assuming any distribution for the noise. But first, let us introduce the problem more formally.

A simple model for the aircraft acoustic signal, as heard by a stationary observer, can be expressed as

$$Z_t = Ae^{j\phi_t} + U_t, \quad t \in \mathbb{R} \tag{5.2}$$

where U_t is a continuous time, zero-mean complex white noise process with variance σ^2, A is a constant assumed herein, without loss of generality, to be unity, and ϕ_t is given by

$$\phi_t = 2\pi \frac{f_a c^2}{c^2 - v^2} \left(t - \sqrt{\frac{h^2 c + v^2 t^2 c + 2v^2 th)}{c^3}} \right) + \phi_0, \quad t \subset \mathbb{R} \tag{5.3}$$

where f_a is the source acoustic frequency, c is the speed of sound in the medium, v is the constant velocity of the aircraft, h is the constant altitude of the aircraft and ϕ_0 is an initial phase constant. From (5.3), the IF, relative to the stationary observer, can be expressed as

$$
\begin{aligned}
f_t &= \frac{1}{2\pi} \frac{d\phi_t}{dt} \\
&= \frac{f_a c^2}{c^2 - v^2} \left(1 - \frac{v^2(t + h/c)}{\sqrt{h^2(c^2 - v^2) + v^2 c^2(t + h/c)^2}} \right), \quad t \in \mathbb{R} \tag{5.4}
\end{aligned}
$$

For a given f_t or ϕ_t, and c, the aircraft parameters collected in the vector $\boldsymbol{\theta} = (f_a, h, v, t_0)'$ can be uniquely determined from the phase model (5.3) or observer IF model (5.4).

To illustrate the use of the bootstrap, we consider two different estimation schemes. The first scheme estimates the observer IF, as modelled by (5.4), using central finite difference (CFD) methods (Boashash, 1992a,b). The second one uses unwrapping of the phase, as modelled by (5.3). The IF and phase estimates are then fitted to the respective models in a least squares sense to provide an estimate of the aircraft parameters. Bootstrap techniques are finally applied to provide confidence bounds for the parameters based on these two estimates. Considering discrete-time signals as appropriate sampled versions of the continuous-time signals, we denote the IF by f_t, the phase by ϕ_t, and the observed signal by Z_t, $t \in \mathbb{Z}$.

The procedures for confidence interval estimation are described in Tables 5.1 and 5.2 for estimation of the aircraft parameters based on the CFD and phase unwrapping respectively.

The methods have been extensively tested with simulated data to check their applicability. Comparisons with Monte Carlo simulations showed their excellent performance (Reid *et al.*, 1996; Reid, 1997; Zoubir and Boashash, 1998). In this contribution, we will only report results with real data.

5.2.2 Results with real passive acoustic data

We will restrict our analysis to one estimation technique, namely phase unwrapping. We will only show one example from a chartered aircraft that flew five times over an observer position as indicated in Figure 5.7. The aircraft was equipped with a differential GPS to record the flight parameters during each of the fly-overs. An example of an estimate of the Doppler profile (instantaneous frequency) is given in Figure 5.8.

The data from three runs and the estimated confidence bands with the bootstrap, following the procedure of Table 5.2, are collected in Table 5.3.

We note that the variation in the interval lengths of each of the parameters in Table 5.3 reflects the sensitivity of the parameters in the observer models to the form of the estimated phase. The above results show, however, that the bootstrap provides a practical flight parameter estimation scheme. It does not assume any statistical distribution of the parameter estimates and can be applied in the absence of multiple acoustic realisations, which are unavailable in this application. The results based on the procedure of Table 5.1 are acceptable, however, the confidence bounds based on unwrapping the phase are much tighter than those of the CFD based estimates.

Table 5.1. *Bootstrap procedure for a confidence interval of flight parameter estimators based on the CFD estimate.*

Step 0. Collect and sample the data to obtain Z_t, $t = -\frac{n}{2}, \ldots, \frac{n}{2} - 1$.

Step 1. Unwrap the phase of the n point signal Z_t to provide a non-decreasing function $\hat{\phi}_t$ which approximates the true phase ϕ_t (Boashash *et al.*, 1991).

Step 2. Estimate the IF by

$$\hat{f}_t = \frac{1}{4\pi}\left(\hat{\phi}_{t+1} - \hat{\phi}_{t-1}\right), \quad t = -n/2 + 1, \ldots, n/2 - 2. \quad (5.5)$$

Step 3. Obtain $\hat{\boldsymbol{\theta}}$, an initial estimate of the aircraft parameter vector $\boldsymbol{\theta}$ by fitting the non-linear IF model $f_{t;\boldsymbol{\theta}}$ to $\hat{f}_t$ in a least squares sense.

Step 4. Consider only every third element of $\hat{f}_t$ and every third element of $f_{t;\hat{\boldsymbol{\theta}}}$

$$\tilde{f}_t = \hat{f}_{3t}, \qquad f_{t;\hat{\boldsymbol{\theta}}}^{(d)} = f_{3t;\hat{\boldsymbol{\theta}}}, \qquad t = -n/6 + 1, \ldots, n/6 - 2,$$

where for convenience it is assumed that $n/6 \in \mathbb{Z}^+$.

Step 5. Compute the residuals $\hat{\varepsilon}_t = f_{t;\hat{\boldsymbol{\theta}}}^{(d)} - \tilde{f}_t$, $\quad t = -n/6 + 1, \ldots, n/6 - 2$.

Step 6. Compute $\hat{\sigma}_i$, a bootstrap estimate of the standard deviation of $\hat{\theta}_i$, $i = 1, \ldots, 4$.

Step 7. Draw a random sample $\mathcal{X}^* = \{\hat{\varepsilon}_{-n/6+1}^*, \ldots, \hat{\varepsilon}_{n/6-2}^*\}$, with replacement, from $\mathcal{X} = \{\hat{\varepsilon}_{-n/6+1}, \ldots, \hat{\varepsilon}_{n/6-2}\}$ and construct

$$\hat{f}_t^* = f_{t;\hat{\boldsymbol{\theta}}}^{(d)} + \hat{\varepsilon}_t^*.$$

Step 8. Obtain and record the bootstrap estimates of the aircraft parameters $\hat{\boldsymbol{\theta}}^*$ by fitting the observer frequency model to $\hat{f}_t^*$ in a least squares sense.

Step 9. Estimate the standard deviation of $\hat{\theta}_i^*$ using a double step bootstrap and compute and record the bootstrap statistics

$$T_{n,i}^* = \frac{\hat{\theta}_i^* - \hat{\theta}_i}{\hat{\sigma}_i^*}, \qquad i = 1, \ldots, 4.$$

Step 10. Repeat Steps 7 through 9 a large number of times B.

Step 11. For each parameter, order the bootstrap estimates as $T_{n,i}^{*(1)} \leq T_{n,i}^{*(2)} \leq \ldots \leq T_{n,i}^{*(B)}$ and compute the $100(1 - \alpha)\%$ confidence interval as

$$\left(T_{n,i}^{*(U)}\hat{\sigma}_i + \hat{\theta}_i, \ T_{n,i}^{*(L)}\hat{\sigma}_i + \hat{\theta}_i\right)$$

where $U = B - \lfloor B\alpha/2 \rfloor + 1$ and $L = \lfloor B\alpha/2 \rfloor$.

Table 5.2. *Bootstrap procedure for a confidence interval of flight parameter estimators based on the unwrapped phase estimate.*

Step 0. Collect and sample the data to obtain Z_t, $t = -\frac{n}{2}, \dots, \frac{n}{2} - 1$.

Step 1. Unwrap the phase of the n point signal Z_t to provide a non-decreasing function $\hat{\phi}_t$ which approximates the true phase ϕ_t.

Step 2. Obtain $\hat{\boldsymbol{\theta}}$, an initial estimate of the aircraft parameters by fitting the non-linear observer phase model (5.3) $\phi_{t;\boldsymbol{\theta}}$ to $\hat{\phi}_t$ in a least squares sense.

Step 3. Compute the residuals

$$\hat{\varepsilon}_t = \phi_{t;\hat{\boldsymbol{\theta}}} - \hat{\phi}_t, \qquad t = -n/2, \dots, n/2 - 1.$$

Step 4. Compute $\hat{\sigma}_i$, a bootstrap estimate of the standard deviation of $\hat{\theta}_i$, $i = 1, \dots, 4$.

Step 5. Draw a random sample $\mathcal{X}^* = \{\hat{\varepsilon}^*_{-n/2}, \dots, \hat{\varepsilon}^*_{n/2-1}\}$, with replacement, from $\mathcal{X} = \{\hat{\varepsilon}_{-n/2}, \dots, \hat{\varepsilon}_{n/2-1}\}$ and construct

$$\hat{\phi}^*_t = \phi_{t;\hat{\boldsymbol{\theta}}} + \hat{\varepsilon}^*_t.$$

Step 6. Obtain and record the bootstrap estimates of the aircraft parameters $\hat{\boldsymbol{\theta}}^*$ by fitting the observer phase model to $\hat{\phi}^*_t$ in a least squares sense.

Step 7. Estimate the standard deviation of $\hat{\theta}^*_i$ using a double step bootstrap and compute and record the bootstrap statistics

$$T^*_{n,i} = \frac{\hat{\theta}^*_i - \hat{\theta}_i}{\hat{\sigma}^*_i}, \qquad i = 1, \dots, 4.$$

Step 8. Repeat Steps 5 through 7 a large number of times B.

Step 9. For each parameter, order the bootstrap estimates as $T^{*(1)}_{n,i} \leq T^{*(2)}_{n,i} \leq \dots \leq T^{*(B)}_{n,i}$ and compute the $100(1-\alpha)\%$ confidence interval as

$$\left(T^{*(U)}_{n,i} \hat{\sigma}_i + \hat{\theta}_i, \quad T^{*(L)}_{n,i} \hat{\sigma}_i + \hat{\theta}_i \right)$$

where $U = B - \lfloor B\alpha/2 \rfloor + 1$ and $L = \lfloor B\alpha/2 \rfloor$.

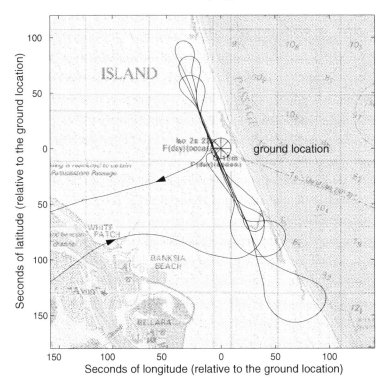

Fig. 5.7. The complete flight path of an aircraft acoustic data experiment conducted at Bribie Island, Queensland, Australia, showing five fly-overs. The ground location of an observer is indicated by the cartwheel symbol.

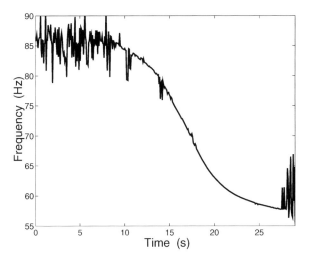

Fig. 5.8. Doppler signature of an acoustic signal for the aircraft scenario.

Table 5.3. *The bootstrap derived 95% confidence bounds for three real test signals using the unwrapped phase based parameter estimate and $B = 1000$.*

	Scenario	h [m]	v [m/s]	t_0 [s]	f_a [Hz]
	Nominal Value	149.05	36.61	0.00	76.91
Run 1.	Upper Bound	142.03	36.50	0.02	76.99
	Lower Bound	138.16	36.07	-0.03	76.92
	Interval length	3.87	0.43	0.05	0.07
	Nominal Value	152.31	52.94	0.00	77.90
Run 2.	Upper Bound	161.22	53.27	-0.04	78.18
	Lower Bound	156.45	52.64	-0.06	78.13
	Interval length	4.77	0.62	0.02	0.05
	Nominal Value	166.52	47.75	0.00	75.94
Run 3.	Upper Bound	232.30	57.29	0.10	76.48
	Lower Bound	193.18	53.47	-0.14	75.87
	Interval length	39.13	3.83	0.24	0.60

5.3 Landmine detection

It is estimated that more than one hundred million active anti-personnel landmines are currently scattered in over 70 countries (Red Cross, 1996). In military applications, ploughs and heavy rollers, which can clear narrow paths through a minefield, can be used. However, their detection rate is not acceptable for civilians. Thus, the clearance of mines becomes a very laborious and dangerous task, especially in post-war regions that can ill-afford to have large regions of unproductive land.

Ground Penetrating Radar (GPR) can detect discontinuities in the electric properties of the soil, and therefore has the potential to detect buried plastic landmines with little metal content. A GPR system is sensitive to changes in the soil conditions, e.g. moisture content and soil type, hence robust methods for detection are required that can adapt to changes in the soil whilst still detecting targets with a high probability.

We are particularly interested in the application of detectors based on bootstrap techniques used in a GPR system. For this we use an FR-127-MSCB Impulse GPR (ImGPR) system developed by the Commonwealth Scientific and Industrial Research Organisation (CSIRO), Australia. The system has 127 soundings per second, each composed of 512 samples with 12 bit accuracy and bistatic bow-tie antennae transmitting wide-band, ultra-short duration pulses. The centre frequency is 1.4 GHz. Three types of surrogate minelike targets (PMN, PMN-2, and M14 AP) with non-metallic casings are used in the experiment (see Table 5.4).

Table 5.4. *Minelike surrogate target used in the study.*

Target	Surrogate	Dimensions diam×height	Orientation
ST-AP(1)	M14	52×42 mm	not critical
ST-AP(2)	PMN	118×50 mm	sensitive
ST-AP(3)	PMN2	115×53 mm	sensitive

It is difficult to define a complete, physically-motivated signal model of the GPR backscatter waveform. For the purpose of detection or classification of landmine targets we usually choose either a parametric approach in which the received signal is modelled by a known function, except for some parameters, or a non-parametric approach in which a certain feature of the received signal is exploited. The received signal of a GPR is usually assumed to be comprised of four major components (Daniels *et al.*, 1988):

(i) the background signal – always present and varying with the environment,

(ii) the target signal – not necessarily present,

(iii) the antenna cross-talk signal – relatively small for a crossed dipole antenna, and

(iv) the measurement noise – assumed to be spatially and temporarily white Gaussian.

Extensive analysis of real data has shown that a polynomial amplitude–polynomial phase model is suitable for modelling the GPR signal, i.e.

$$s_t = \left[\sum_{n=0}^{P_a} a_n t^n \right] \exp \left[j \sum_{m=0}^{P_b} b_m t^m \right] + z_t \qquad t = 1, \ldots, n, \qquad (5.6)$$

where z_t, $t = 1, \ldots, n$ are realisations of stationary interference and the amplitude and frequency modulation are described by polynomials of order P_a and P_b respectively.

We will report here only one experiment with a surrogate and a stainless steel target. For more details, the reader is referred to earlier works (Barkat *et al.*, 2000; Iskander *et al.*, 2000b; Zoubir *et al.*, 1999, 2002). We have applied the signal model (5.6) for GPR returns from clay soil, containing a small plastic target, denoted ST-AP(1), which is a surrogate (replica) for the M14 anti-personnel mine. The ST-AP(1) target has a PVC casing which is filled with paraffin wax. A solid stainless steel cylinder 5 cm in diameter and length, denoted by SS05x05, is also present in the soil. Phase and amplitude parameters are estimated using the discrete polynomial phase transform (Peleg and Friedlander, 1995) and the method of least squares respectively. An example of the model fit to the real GPR signal is shown in Figure 5.9. After performing model fit to real GPR data, it has been observed that the means of the coefficient distributions change in the presence of a target as indicated in Figure 5.10 for a second order fit.

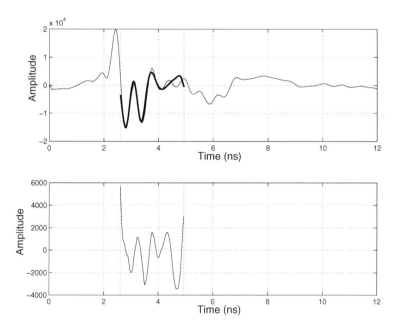

Fig. 5.9. The fit of the polynomial amplitude–polynomial phase model (bold line) to GPR data (fine line) from non-homogeneous soil (top). Residual of the fit to data (bottom).

The outcomes of the experimental work led to a detection scheme based on the coefficients of the polynomial phase of the signal. Consider the following example. For the data set shown in Figure 5.10, b_2 is seen to decrease in the presence of a target, while in some cases, particularly when the soil type is loam, b_2 *increases* with the presence of a target. This suggests the test of the hypothesis-alternative pair (H, K),

$$H \quad : \quad b_2 = b_{2,0}$$
$$K \quad : \quad b_2 \neq b_{2,0}$$

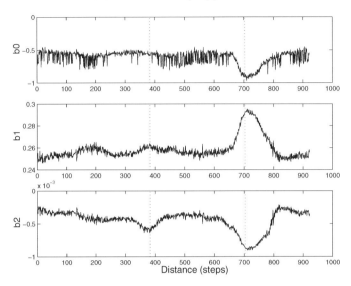

Fig. 5.10. Phase coefficients from an order 2 model for GPR data. Vertical dotted lines indicate the approximate position of ST-AP(1) (left) and SS05x05 (right) targets.

The test statistic is then

$$T_n = \frac{\hat{b}_2 - b_{2,0}}{\hat{\sigma}_{\hat{b}_2}}$$

where $\hat{\sigma}_{\hat{b}_2}$ is an estimate of the standard deviation of $\hat{b}_2$ and $b_{2,0}$ is a nominal value which can be selected based on target-free measurements. Since the exact distribution of $\hat{b}_2$ is unknown and has unknown variance, we use the bootstrap to determine thresholds as shown in Table 5.5.

One of the results of the bootstrap based detection scheme is shown in Figure 5.11, where a scan of the two surrogate targets mentioned above is shown together with the corresponding test statistics and the detection decision.

This example shows another useful and powerful application of the bootstrap. Overall, the results were encouraging. The targets have been correctly detected, while false alarms appear to be concentrated in areas near the targets – rather than *true* false alarms that occur far from the targets in background only areas. These near-target false alarms may be triggered by disturbances to the soil structure caused by the burying of the target.

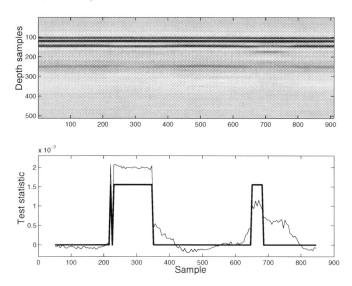

Fig. 5.11. The *B*-scan with ST-AP(1) and SS05x05 targets buried at 5 cm in clay (top) and the corresponding test statistics (thin line) and detection decision (thick line) (bottom).

5.4 Noise floor estimation in over-the-horizon radar

The propagation environment of a high frequency over-the-horizon radar (see Figure 5.12) is highly non-stationary (Barrick and Snider, 1977). Time-varying propagation conditions and fluctuating HF noise characteristics dictate that radar system parameters should continually be revised by the radar operator if optimum performance is to be achieved. In the typical operation of a skywave radar (Anderson, 1986), several performance statistics are obtained from the received radar signals and are provided to the operator to guide the radar optimisation process.

One such radar performance statistic is the noise floor. The noise floor estimate is computed by taking a trimmed mean of the power estimates in the Doppler spectrum after excluding the low frequency Doppler energy associated with any stationary targets or ground clutter. The noise floor is used directly by the operator and also as a component in other radar performance statistics. The noise floor estimate alone is insufficient. The radar operator requires supplementary information describing the accuracy of the computed noise floor, i.e., a confidence measure. The limited number of samples available to the radar operator and the non-Gaussian nature of the noise field make it particularly difficult to estimate a confidence measure

Table 5.5. *The algorithm for the bootstrap based landmine detection using GPR.*

Step 1. s_t, $t = 1, ..., n$, is the sampled GPR return signal. Fit a polynomial amplitude–polynomial phase model to the data. From the model, find an estimate of the second order phase parameter, $\hat{b}_2$.
Step 2. Form the residual signal $\hat{r}_t = s_t - \hat{g}_t$, where $\hat{g}_t$ is the polynomial amplitude–polynomial phase model corresponding to estimated parameters.
Step 3. Whiten $\hat{r}_t$ by removing an AR model of suitable order to obtain the innovations $\hat{z}_t$.
Step 4. Resample from $\hat{z}_t$ using the block of blocks bootstrap to obtain $\hat{z}_t^*$.
Step 5. Repeat Step 4 B times to obtain $\hat{z}_t^{*1}, ..., \hat{z}_t^{*B}$.
Step 6. Generate B bootstrap residual signals $\hat{r}_t^{*i}$, for $i = 1, ..., B$ by filtering $\hat{z}_t^{*i}$ with the AR process obtained in (5.5).
Step 7. Generate B bootstrap signals $\hat{g}_t^{*i} = \hat{g}_t + \hat{r}_t^{*i}$, for $i = 1, ..., B$.
Step 8. Estimate the second phase coefficient from $\hat{g}_t^{*i}$ to obtain $\hat{b}_2^{*i}$ for $i = 1, ..., B$.
Step 9. Calculate the bootstrap statistics $T_n^{*i} = \frac{\hat{b}_2^{*i} - \hat{b}_2}{\hat{\sigma}_{\hat{b}_2}^*}$ for $i = 1, ..., B$.
Step 10. Compare the test statistic $T_n = \frac{\hat{b}_2 - b_{2,0}}{\hat{\sigma}_{\hat{b}_2}}$ to the empirical distribution of T_n^*. Reject H if T_n is in the tail of the distribution, otherwise retain H.

reliably. Accuracy measures may also be developed for other performance statistics. The bootstrap is an ideal tool in such conditions.

5.4.1 Principle of the trimmed mean

Consider the following estimation problem. Let θ be a characteristic of F_θ and suppose we wish to choose an estimate from a class of estimates $\{\hat{\theta}(\gamma) : \gamma \in \mathcal{A}\}$ indexed by a parameter γ, which is usually selected according to some optimality criterion. If the estimates are unbiased, it is reasonable to use the estimate that leads to the minimal variance. However, the variance of the estimate $\hat{\theta}(\gamma)$ may depend on unknown characteristics. Thus the optimal parameter γ_o may be unknown. This problem is encountered with the trimmed mean, where one wants to optimise the amount of trimming. Thus, the objective is to find γ by estimating the unknown variance of

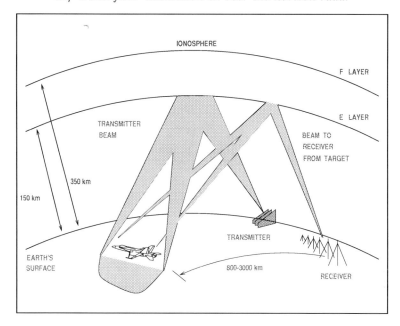

Fig. 5.12. The over-the-horizon-radar scenario.

the estimates and select a parameter value, say $\hat{\theta}(\gamma)$, which minimises the variance of the estimator.

This problem has been solved theoretically by Jaeckel (1971) for a symmetric distribution F_θ. He proposes an estimate of the asymptotic variance of the trimmed mean and verifies that the resulting randomly trimmed mean has the same asymptotic distribution as the trimmed mean with the smallest asymptotic variance. However, the trim is restricted to the interval $[0, 25]\%$ (Mehrotra *et al.*, 1991). In addition to the fact that for a small sample size asymptotic results are invalid, explicit expressions for the asymptotic variance are not available outside the interval $[0, 25]\%$.

Let $\mathcal{X} = \{X_1, \ldots, X_n\}$ be a random sample drawn from a distribution function F which is symmetric about zero and let $X_{(1)}, \ldots, X_{(n)}$ denote the order statistic. For an integer γ less than $n/2$, the γ-trimmed mean of the sample $\mathcal{X}$ is given by

$$\hat{\mu}(\gamma) = \frac{1}{n - 2\gamma} \sum_{i=\gamma+1}^{n-\gamma} X_{(i)}. \tag{5.7}$$

The aim is to find γ such that $\hat{\mu}(\gamma)$ has the smallest possible variance. To select the best $\gamma = \gamma_o \in \mathcal{A}$, Jaeckel (1971) proposed using the trimmed mean whose asymptotic variance is minimum. He estimated the asymptotic variance

$$\sigma(\gamma)^2 = \frac{1}{(1-2\gamma)^2} \left(\int_{\zeta_\gamma}^{\zeta_{1-\gamma}} x^2 dF(x) + 2\gamma\zeta_\gamma^2 \right), \tag{5.8}$$

where $\zeta_\gamma = F^{-1}(\gamma)$ and $\zeta_{1-\gamma} = F^{-1}(1-\gamma)$, by its sample analogue

$$s(\gamma)^2 = \frac{1}{(1-2\gamma)^2} \left(\frac{1}{n} \sum_{i=\gamma+1}^{n-\gamma} \tilde{X}_{i,\gamma}^2 + \gamma\tilde{X}_{\gamma+1,\gamma}^2 + \gamma\tilde{X}_{n-\gamma,\gamma}^2 \right), \tag{5.9}$$

where $\tilde{X}_{i,\gamma} = X_{(i)} - \hat{\mu}(\gamma)$, $i = 1, \ldots, n$. Jaeckel proposed then the estimate $\hat{\mu}(\hat{\gamma})$, where $\hat{\gamma}$ minimises the estimated variance s^2 over a subset of $[0, 0.5)$.

In other problems, however, explicit expressions for the asymptotic variance may not be available. This includes the case where F_θ is an asymmetric distribution. Moreover, for a small sample size the asymptotic result may be invalid. To avoid this difficulty the following bootstrap based procedure is proposed.

5.4.2 Optimal trimming

In order to find an optimal value of the trim we estimate the variances of the trimmed means using the bootstrap and then select the trimming portion which minimises the bootstrap variance estimate (Léger and Romano, 1990; Mehrotra *et al.*, 1991).

Let us consider first a more general case where the symmetric assumption is dropped. Then, the trimmed mean will estimate some quantity depending on the amount of left and right trimming. For integers $\gamma < n/2$ and $\kappa < n/2$, the $\gamma\kappa$-trimmed mean based on the sample $\mathcal{X}$ is given by

$$\hat{\mu}(\gamma, \kappa) = \frac{1}{n - \gamma - \kappa} \sum_{i=\gamma+1}^{n-\kappa} X_{(i)}, \quad \gamma, \kappa < n/2. \tag{5.10}$$

The procedure for finding the optimal trim when the data of the sample $\mathcal{X}$ are independent and identically distributed is given in Table 5.6.

The optimal estimator $\hat{\mu}(\gamma_o, \kappa_o) = \hat{\mu}_o$ can be further refined using the so-called augmented sample. This is achieved by augmenting the sample $\mathcal{X}$ by the values $2\hat{\mu}_o - X_1, \ldots, 2\hat{\mu}_o - X_n$ which results in a new sample

$$\mathcal{X}_A = (X_1, \ldots, X_n, 2\hat{\mu}_o - X_1, \ldots, 2\hat{\mu}_o - X_n).$$

In order to find the optimal trim, the procedure outlined in Table 5.6 is then repeated for the augmented sample $\mathcal{X}_A$ (Mehrotra *et al.*, 1991).

Table 5.6. *Bootstrap procedure for the optimal trimmed mean for iid data.*

Step 1. *Initial conditions.* Select the initial trim $\gamma = 0$ and $\kappa = 0$.

Step 2. *Resampling.* Draw a large number, say B, of independent samples

$$\mathcal{X}_1^* = \{X_{11}^*, \ldots, X_{1n}^*\}, \ldots, \mathcal{X}_B^* = \{X_{B1}^*, \ldots, X_{Bn}^*\}$$

of size n from $\mathcal{X}$.

Step 3. *Calculation of the bootstrap statistic.* For each bootstrap sample $\mathcal{X}_j^*$, $j = 1, \ldots, B$, calculate the trimmed mean

$$\hat{\mu}_j^*(\gamma, \kappa) = \frac{1}{n - \gamma - \kappa} \sum_{i=\gamma+1}^{n-\kappa} X_{j(i)}^*, \qquad j = 1, \ldots, B.$$

Step 4. *Variance estimation.* Using bootstrap based trimmed mean values, calculate the estimate of the variance

$$\hat{\sigma}^{*2} = \frac{1}{B-1} \sum_{i=1}^{B} \left(\hat{\mu}_i^*(\gamma, \kappa) - \frac{1}{B} \sum_{j=1}^{B} \hat{\mu}_j^*(\gamma, \kappa) \right)^2.$$

Step 5. *Repetition.* Repeat Steps 2–4 using different combinations of the trim γ and κ.

Step 6. *Optimal trim.* Choose the setting of the trim that results in a minimal variance estimate.

5.4.3 Noise floor estimation

As mentioned earlier, the noise floor in a radar system can be estimated by a trimmed mean of the Doppler spectrum. The bootstrap procedure for a trimmed mean given in Table 5.6 can be used only in the case of iid data and is inapplicable here. To cater for dependent data we employ the residual based approach outlined earlier in Section 2.1.4 (see Example 2.1.10). Thus, we estimate residuals for the Doppler spectrum by taking the ratio of the periodogram and a spectral estimate of the raw data. The residuals are used for resampling to generate bootstrap spectral estimates.

We proceed then by replacing X_i in Table 5.6 by $\hat{C}_{XX}(\omega_i)$, where $\omega_i = 2\pi i/n$, $i = 1, \ldots, n$, are discrete frequencies and n is the number of observations. The procedure makes the assumption that the residuals are iid for distinct discrete frequencies. The optimal noise floor is found by minimising the bootstrap variance estimate of the trimmed spectrum with respect to (γ, κ). The adapted bootstrap algorithm for choosing the optimal trim for the noise floor estimator is shown in Table 5.7.

Example 5.4.1 Noise floor estimation.

We first run a simulation to show optimal estimation of the noise floor using the bootstrapped trimmed mean. We first simulate the target scenario using a complex-valued AR(3) process with coefficients $1, -0.8334 + 0.6662j, -1.0244 - 0.4031j, 0.9036 - 0.2758j$. Then we filter the clutter concentrated around zero-Doppler and calculate, using the procedure of Table 5.7, the optimal trimmed mean of the noise floor. The complete code for this example is included in Section A1.9 of Appendix 1. The result of the simulation is shown in Figure 5.13 while a result for real over-the-horizon radar (OTHR) data is given in Figures 5.14 and 5.15. These results indicate that the method gives a reasonable estimate of the noise floor.

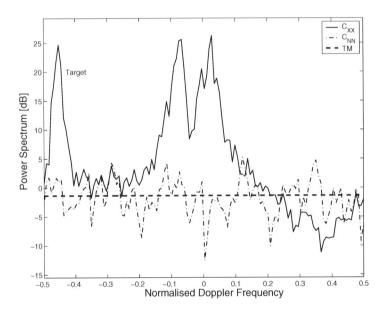

Fig. 5.13. Spectrum and estimated noise floor for an AR(3) model.

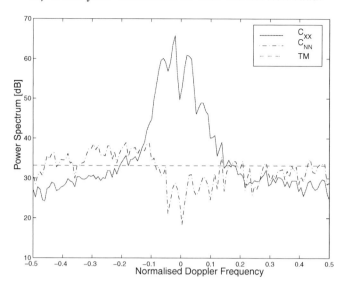

Fig. 5.14. Doppler spectrum and estimated noise floor for a real over-the-horizon radar return (no target present), $B = 100$, $\hat{\mu}(\hat{\gamma}, \hat{\kappa}) = 2.0738 \times 10^3$.

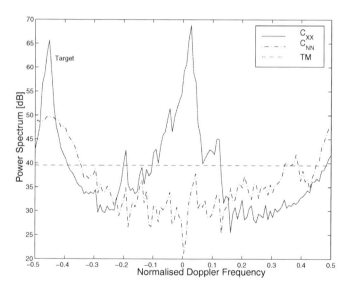

Fig. 5.15. Doppler spectrum and estimated noise floor for a real over-the-horizon radar return (target present), $B = 100$, $\hat{\theta}(\hat{\gamma}, \hat{\kappa}) = 8.9513 \times 10^3$.

Table 5.7. *Bootstrap procedure for finding the optimal trimmed mean for noise floor estimation.*

Step 1. *Initial conditions.* Select the initial trim $\gamma = \kappa = 0$.

Step 2. *Resampling.* Draw a large number, say B, of independent samples

$$\mathcal{X}_1^* = \{\hat{C}_{XX}^{*(1)}(\omega_1), \ldots, \hat{C}_{XX}^{*(1)}(\omega_M)\}$$

$$\ldots$$

$$\mathcal{X}_B^* = \{\hat{C}_{XX}^{*(B)}(\omega_1), \ldots, \hat{C}_{XX}^{*(B)}(\omega_M)\}$$

Step 3. *Calculation of the trimmed mean.* For each bootstrap sample $\mathcal{X}_j^*$, $j = 1, \ldots, B$, calculate the trimmed mean

$$\hat{\mu}_j^*(\gamma, \kappa) = \frac{1}{M - \gamma - \kappa} \sum_{i=\gamma+1}^{n-\kappa} \hat{C}_{XX}^{*(j)}(\omega_i), \qquad j = 1, \ldots, B.$$

Step 4. *Variance estimation.* Calculate the estimate of the variance

$$\hat{\sigma}^{*2} = \frac{1}{B-1} \sum_{i=1}^{B} \left(\hat{\mu}_i^*(\gamma, \kappa) - \frac{1}{B} \sum_{j=1}^{B} \hat{\mu}_j^*(\gamma, \kappa) \right)^2.$$

Step 5. *Repetition.* Repeat Steps 2–4 using different combinations of the trim γ and κ.

Step 6. *Optimal trim.* Choose the setting of the trim that results in a minimal variance estimate.

5.5 Model order selection for corneal elevation

The cornea is the major refracting component of the human eye. It contributes approximately 60% of the eye's total optical power. Ideally, the anterior surface of the cornea would be a prolate ellipsoid. However, in nature this is not the case and a variety of corneal shapes is possible. Thus, optometrists, ophthalmologists and vision scientists are interested in modelling corneal elevation and finding deviations of the cornea from the ideal shape. This deviation can be related to the so-called wavefront aberration which determines vision performance (Iskander *et al.*, 2000a).

Videokeratoscopy is the current standard in accurate measurement of corneal elevation. The majority of current videokeratoscopes are based on a placido disk approach (Mandell, 1996). The placido disk forms the target of rings, while a digital camera captures the reflected image from the tear film.

The acquired image is processed so that the topographical corneal height can be calculated.

Several techniques exist for modelling corneal surfaces from videokerato-graphic data. The most popular include Zernike polynomial expansion and other radial polynomial expansions (Iskander *et al.*, 2001, 2002).

We model the corneal surface by a finite series of Zernike polynomials

$$C(\rho, \theta) = \sum_{p=1}^{P} a_p Z_p(\rho, \theta) + \varepsilon, \tag{5.11}$$

where $C(\rho, \theta)$ denotes the corneal surface, $Z_p(\rho, \theta)$, $p = 1, \ldots, P$, is the pth Zernike polynomial, a_p, $p = 1, \ldots, P$, is the coefficient associated with $Z_p(\rho, \theta)$, P is the order, ρ is the normalised distance from the origin, θ is the angle, and ε represents measurement and modelling error (noise). We assume that the noise is iid with zero mean and finite variance. We consider the general case where the distribution of the additive noise is unknown.

The pth-order Zernike polynomial is defined as (Noll, 1976)

$$Z_p(\rho, \theta) = \begin{cases} \sqrt{2(n+1)} R_n^m(\rho) \cos(m\theta), \text{ even } p, m \neq 0 \\ \sqrt{2(n+1)} R_n^m(\rho) \sin(m\theta), \text{ odd } p, m \neq 0 \\ \sqrt{n+1} R_n^0(\rho), \qquad\qquad m = 0 \end{cases}$$

where n is the radial degree, m is the azimuthal frequency, and

$$R_n^m(\rho) = \sum_{s=0}^{(n-m)/2} \frac{(-1)^s (n-s)!}{s! \left(\frac{n+m}{2} - s\right)! \left(\frac{n-m}{2} - s\right)!} \rho^{n-2s}.$$

The radial degree and the azimuthal frequency are integers which satisfy $m \leq n$ and $n - |m| =$ even. The radial degree n and the azimuthal frequency m can be evaluated from the polynomial-ordering number p using $n = \lfloor q_1 \rfloor - 1$, and $m = q_2 + (n + (q_2 \bmod 2) \bmod 2)$ respectively, where $q_1 = 0.5(1 + \sqrt{8p - 7})$, $p = 1, \ldots, P$, $q_2 = \lfloor (n+1)(q_1 - n - 1) \rfloor$, $\lfloor \cdot \rfloor$ is the floor operator and mod denotes the modulus operator.

The problem of finding an appropriate set of Zernike polynomials that would best suit a videokeratoscopic measurement is a classical model selection problem as outlined in Chapter 4. We note that in practice it is sufficient to choose only the model order of the polynomial expansion, P, rather than a particular model being a subset of $\{Z_1(\rho, \theta), \ldots, Z_P(\rho, \theta)\}$. A typical raw videokeratoscopic measurement of the corneal elevation is shown in Figure 5.16.

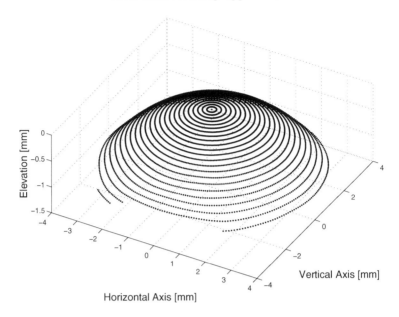

Fig. 5.16. Raw elevation videokeratoscopic data for an 8 mm corneal diameter.

The Zernike polynomials are orthogonal in a unit circle. However, in a discrete case where the number of samples is not large, further orthogonalisation of the function is performed using a Gram–Schmidt procedure. After orthogonalisation we can form the following linear (in the parameters) model

$$\boldsymbol{C} = \boldsymbol{Z}\boldsymbol{a} + \varepsilon, \qquad (5.12)$$

where $\boldsymbol{C}$ is a D-element column vector of corneal surface evaluated at discrete points (ρ_d, θ_d), $d = 1, \ldots, D$, $\boldsymbol{Z}$ is a $(D \times P)$ matrix of discrete orthogonalised Zernike polynomials $Z_p(\rho_d, \theta_d)$, $\boldsymbol{a}$ is a P-element column vector of Zernike coefficients, and ε represents a D-element column vector of the measurement and modelling error.

A bootstrap procedure for finding the optimal Zernike fit to the corneal elevation is shown in Table 5.8.

We have experimented with a large number of human corneas. Normal as well as pathological cases have been considered. We showed that for normal corneas, the optimal number of Zernike terms usually corresponds to the fourth or fifth radial order expansion of Zernike polynomials. On the other hand, for distorted corneas such as those encountered in keratoconus or in surgically altered cases, the estimated model is up to several radial orders higher than for normal corneas. These results are in agreement with recently

Table 5.8. *Bootstrap procedure for model order selection of a Zernike polynomial expansion.*

Step 0. *Experiment:* Measure the anterior surface of the eye using a videokeratoscope and collect the elevation data into a column vector.

Step 1. *Initial estimation:* Select $\beta = \beta_{\max}$, find the estimate $\hat{a}_\beta$ of a_β using the method of least squares and compute $\hat{C}(\rho_d, \theta_d) = \sum_{p=1}^{\beta_{\max}} \hat{a}_p Z_p(\rho_d, \theta_d)$.

Step 2. *Compute residuals:* Calculate the residuals $\hat{r}_d = C(\rho_d, \theta_d) - \hat{C}(\rho_d, \theta_d)$, $d = 1, \ldots, D$.

Step 3. *Scaling:* Rescale the empirical residuals
$$\tilde{r}_d = \sqrt{D/L_D}\left(\hat{r}_d - D^{-1}\sum_{d=1}^{D}\hat{r}_d\right)$$

Step 4. *Try every model:* For all $1 \leq \beta \leq \beta_{\max}$

   (a) Calculate $\hat{a}_\beta$ and $\hat{C}(\rho_d, \theta_d)$ as in Step 1.

   (b) Draw independent bootstrap residuals $\tilde{r}_d^*$ with replacement, from the empirical distribution of $\tilde{r}_d$.

   (c) Define the bootstrap surface
$$C^*(\rho_d, \theta_d) = \hat{C}(\rho_d, \theta_d) + \tilde{r}_d^*.$$

   (d) Using $C^*(\rho_d, \theta_d)$ as the *new* surface, compute the least squares estimate of a_β, $\hat{a}_\beta^*$, and calculate $\hat{C}^*(\rho_d, \theta_d) = \sum_{p=1}^{\beta} \hat{a}_p^* Z_p(\rho_d, \theta_d)$ and the sum of the squared errors

$$\mathsf{SSE}^*_{D,L_D}(\beta) = \frac{1}{D}\sum_{d=0}^{D}\left(C(\rho_d, \theta_d) - \hat{C}^*(\rho_d, \theta_d)\right)^2.$$

   (e) Repeat Steps (b)–(d) a large number of times (e.g. 100) to obtain a total of B bootstrap statistics $\mathsf{SSE}^{*,1}_{D,L_D}(\beta), \ldots, \mathsf{SSE}^{*,B}_{D,L_D}(\beta)$, and estimate the bootstrap mean-squared error

$$\bar{\Gamma}_{D,L_D}(\beta) = \frac{1}{B}\sum_{b=1}^{B}\mathsf{SSE}^{*,b}_{D,L_D}(\beta).$$

Step 5. *Model selection:* Choose β for which $\bar{\Gamma}_{D,L_D}(\beta)$ is a minimum.

---

reported independent studies involving ad hoc threshold-based techniques, proving the bootstrap technique is effective in accurately modelling a large variety of corneal shapes.

5.6 Summary

We have reported some applications of the bootstrap to solving signal processing problems. We have first discussed the problem of optimal sensor placement on a spark ignition engine for the detection of knock. We have shown that the bootstrap is a valuable tool to approximate the distributions of statistics required by a hypothesis test. We have then turned our attention to a passive acoustic emission problem where we have estimated confidence intervals for an aircraft's flight parameters. Analytical solutions to this problem are almost impossible due to the nature of the estimation procedure for the parameters of interest. With the bootstrap we have been able to answer the question of confidence interval estimation, with no assumptions on the noise distribution.

We also have looked into the important problem of civilian landmine detection. We have suggested an approach to detect buried landmines using a ground penetrating radar. Preliminary results have shown that the method has its merits, but much more has to be done to solve this very difficult problem. We note that GPR is only one approach to the global problem and many more solutions are being investigated. We have considered another radar application where we have considered noise floor estimation in high frequency over-the-horizon radar. We have devised a bootstrap procedure to optimally estimate the noise floor. The chapter concludes with the estimation of the optimal model for the corneal elevation in the human eye. All results reported in this chapter are based on real data.

Appendix 1
MATLAB codes for the examples

A1.1 Basic non-parametric bootstrap estimation
Example 2.1.3. Bias estimation.

```
n=5;
B=1000;
MC=1000;
X=randn(1,n);
sigma2_u=var(X);
sigma2_b=var(X,1);
for m=1:MC,
    X_star=bootrsp(X,B);
    sigma2_star_u=var(X_star);
    sigma2_star_b=var(X_star,1);
    bias_u(m)=mean(sigma2_star_u-sigma2_b);
    bias_b(m)=mean(sigma2_star_b-sigma2_b);
end
```

Example 2.1.4. Variance estimation.

```
n=50;
B=25;
X=10+5*randn(1,n);
X_star=bootrsp(X,B);
theta_star=mean(X_star);
sigma2_Boot=var(theta_star);
```

Example 2.1.5. Confidence interval for the mean.

```
X=[-2.41 4.86 6.06 9.11 10.20 12.81 13.17 14.10 ...
    15.77 15.79];
B=1000;
alpha=0.05;
X_star=bootrsp(X,B);
mu_star=mean(X_star);
mu_sort=sort(mu_star);
q1=floor(0.5*B*alpha);
q2=N-q1+1;
conf_inter=[mu_sort(q1),mu_sort(q2)];
```

A1.2 The parametric bootstrap

Example 2.1.6. Confidence interval for the mean (a parametric approach).

```
X=[-2.41 4.86 6.06 9.11 10.20 12.81 13.17 14.10 ...
    15.77 15.79];
hat_mu=mean(X);
hat_sigma=std(X);
n=10;
B=1000;
alpha=0.05;
X_star=hat_mu+hat_sigma*randn(n,B);
mu_star=mean(X_star);
mu_sort=sort(mu_star);
q1=floor(0.5*B*alpha);
q2=B-q1+1;
conf_inter=[mu_sort(q1),mu_sort(q2)];
```

A1.3 Bootstrap resampling for dependent data

Example 2.1.9. Variance estimation in AR models.

```
n=128;
B=1000;
a=-0.6;
Zt=randn(1,n);
X=filter(1,[1 a],Zt);
X=X-mean(X);
c0=mean(X.^2);
c1=mean(X(1:n-1).*X(2:n));
ahat=-c1/c0;
zt=filter([1 ahat],1,X);
zt_star=bootrsp(zt,B);
X_star=filter(1,[1 ahat],zt_star);
X_star(:,1)=X(1);
c0_star=mean(X_star.^2);
c1_star=mean(X_star(1:n-1,:).*X_star(2:n,:));
ahat_star=-c1_star./c0_star;
sigma2_star=var(ahat_star);
```

A1.4 The principle of pivoting and variance stabilisation

Example 2.2.1. **Confidence interval for the mean with a pivotal statistic.**

```
X=[-2.41 4.86 6.06 9.11 10.20 12.81 13.17 14.10 ...
    15.77 15.79];
B=1000;
B1=25;
alpha=0.05;
hat_mu=mean(X);
X_star1=bootrsp(X,B1);
mu_star1=mean(X_star1);
hat_sigma=std(mu_star1);
X_star=bootrsp(X,B);
hat_mu_star=mean(X_star);
hat_mu_Y_star=zeros(1,B);
for k=1:N,
    % Nested bootstrap
    X_star_star=bootrsp(X_star(:,k),B1);
```

```
        hat_mu_star_star=mean(X_star_star);
        hat_sigma_star=std(hat_mu_star_star);
        hat_mu_Y_star(k)=(hat_mu_star(k)-hat_mu)/ ...
                           hat_sigma_star;
    end
    mu_Y_sort=sort(hat_mu_Y_star);
    q1=floor(0.5*B*alpha);
    q2=N-q1+1;
    conf_inter=[hat_mu-hat_sigma*mu_Y_sort(q2), ...
                 hat_mu-hat_sigma*mu_Y_sort(q1)];
```

Example. The principle of variance stabilisation.

Below we show a Matlab routine that illustrates how the variance stabilisation is estimated (see Section 2.2). We use the sample mean as a statistic $\hat{\theta}$. The function smooth.m used below is a running line smoother that fits the data by linear least squares as described in Hastie and Tibshirani (1990) (see Appendix 2).

```
    n=50;
    B1=100;
    B2=25;
    B3=1000;
    X=randn(1,n);
    theta=mean(X);
    X_star=bootrsp(X,B1);
    theta_star=mean(X_star);
    for k=1:B1,
        X_star_star=bootrsp(X_star(:,k),B2);
        theta_star2=mean(X_star_star);
        sigmastar2(k)=var(theta_star2);
    end
    [statsort,sigmasort,sigmasm2]=smooth(theta_star, ...
                              sigmastar2,B1/200);
    a=statsort;
    b=sigmasm2.^(-1/2);
    h=zeros(1,B1);
    h(1)=0;
    for i=2:B1,
```

```
    h(i)=h(i-1)+(a(i)-a(i-1))*(b(i)+b(i-1))/2;
end;
```

A1.5 Limitations of bootstrap procedure

Example 2.3.1. A failure of the non-parametric bootstrap.

```
n=50;
B=1000;
X=rand(1,n);
X_sort=sort(X);
hat_theta=X_sort(n);
% Non-parametric bootstrap
X_star=bootrsp(X,B);
X_sort_star=sort(X_star);
hat_theta_star=X_sort_star(n,:);
% Parametric bootstrap
X_star1=rand(n,B)*hat_theta;
X_sort_star1=sort(X_star1);
hat_theta_star1=X_sort_star1(n,:);
```

A1.6 Hypothesis testing

Example 3.4.2. Testing the third-order cumulant using a non-pivoted bootstrap test

Below we show a MATLAB code for the non-pivoted bootstrap test embedded in a Monte Carlo simulation to show that the test maintains the preset level of significance.

```
NoMonteCarlo=1000;
n=100;
B1=1000;
alpha=0.05;
M=floor((B1+1)*(1-alpha));
H=zeros(1,NoMonteCarlo);
```

```
  for ii=1:NoMonteCarlo,
      X=randn(n,1);
      c3_hat=cumulant3(X);
      T_hat=abs(c3_hat-0); %testing for zero
      X_star=bootrsp(X,B1);
      T_star=abs(cumulant3(X_star)-c3_hat); %two sided test
      Tsort=sort(T_star);
      if T_hat>Tsort(M),
          H(ii)=1;
      end
  end
  P_FA=sum(H)/NoMonteCarlo

  P_FA =

          0.0490
```

In Appendix 2, we have included the function boottestnp.m that performs a non-pivotal bootstrap test for an arbitrary test statistic. We can repeat the above simulation using this function in the following manner.

```
  NoMonteCarlo=1000;
  B1=1000;
  n=100;
  alpha=0.05;
  H=zeros(1,NoMonteCarlo);
  for ii=1:NoMonteCarlo,
      X=randn(n,1);
      H(ii)=boottestnp(X,'cumulant3',0,1,alpha,B1);
  end
  P_FA=sum(H)/NoMonteCarlo

  P_FA =

          0.0510
```

The function cumulant3.m simply evaluates the third-order sample cumulant $\hat{c}_3$.

```
function[c3]=cumulant3(x)
%    [c3]=cumulant3(x)
%
```

```
%    An unbiased estimator of the third order cumulant
```

```
[N,M]=size(x);
s1=sum(x);s2=sum(x.^2);s3=sum(x.^3);
c3=1/(N*(N-1)*(N-2)).*(N^2*s3-3*N*s2.*s1+2*s1.^3);
```

Testing the third-order cumulant using a pivotal statistic.

```
  NoMonteCarlo=1000;
  B1=200;
  B2=25;
  n=100;
  alpha=0.05;
  M=floor((B1+1)*(1-alpha));
  H=zeros(1,NoMonteCarlo);
  for ii=1:NoMonteCarlo,
        X=randn(n,1);
        c3_hat=cumulant3(X);
        X_star1=bootrsp(X,B2);            % first boostrap
        sigma=std(cumulant3(X_star1));
        T_hat=abs(c3_hat-0)./sigma;       % testing for zero
        X_star=bootrsp(X,B1);             % main bootstrap
        T_star=zeros(1,B1);
        for jj=1:B1,
            % nested bootstrap
            X_star_star=bootrsp(X_star(:,jj),B2);
            sigma_star=std(cumulant3(X_star_star));
            T_star(jj)=abs(cumulant3(X_star(:,jj))-c3_hat) ...
                    ./sigma_star; % two sided test
        end
        Tsort=sort(T_star);
        if T_hat>Tsort(M),
            H(ii)=1;
        end
  end
  P_FA=sum(H)/NoMonteCarlo

  P_FA =

        0.0500
```

Note that the number of bootstrap resamples in the case of a pivoted statistic is $B_1 \cdot B_2$ (see Section 3.4), because for each bootstrap sample $\mathcal{X}_b^*$, $b = 1, \ldots, B_1$, we need to draw resamples $\mathcal{X}_b^{**}$, $b = 1, \ldots, B_2$.

As in the previous case we can use a general testing function boottest.m that is included in the Appendix 2. We can repeat the above simulation now with this function in the following manner.

```
NoMonteCarlo=1000;
B1=1000;
B2=25;
n=100;
alpha=0.05;
H=zeros(1,NoMonteCarlo);
for ii=1:NoMonteCarlo,
    X=randn(n,1);
    H(ii)=boottest(X,'cumulant3',0,1,alpha,B1,B2);
end
P_FA=sum(H)/NoMonteCarlo


P_FA =

       0.0510
```

Testing the third-order cumulant with variance stabilisation.

In this example, we also illustrate a bootstrap hypothesis test with variance stabilisation. We use the function boottestvs.m from Appendix 2 as follows.

```
NoMonteCarlo=100;
B1=100;
B2=25;
B3=1000;
n=100;
alpha=0.05;
H=zeros(1,NoMonteCarlo);
for ii=1:NoMonteCarlo,
    X=randn(n,1);
    H(ii)=boottestvs(X,'cumulant3',0,1,alpha,B1,B2,B3);
end
P_FA=sum(H)/NoMonteCarlo
```

```
P_FA =

        0.0520
```

A1.7 The bootstrap matched filter

Example 3.6.1. Performance of the matched filter: Gaussian noise.

```
NoMonteCarlo=100;
n=10;
t=0:n-1;
SNR=7;
A=sqrt(20^(SNR/10));
st=(A*sin(2*pi*t/6))';
B1=1000;
B2=25;
alpha=0.05;
P_FA_data=zeros(4,NoMonteCarlo);
P_D_data=zeros(4,NoMonteCarlo);
for ii=1:NoMonteCarlo,
    Wt=randn(n,1);
    P_FA_data(:,ii)=bootmf(Wt,st,alpha,B1,B2);
    P_D_data(:,ii)=bootmf(st+Wt,st,alpha,B1,B2);
end
P_FA=sum(P_FA_data')/NoMonteCarlo;
P_D=sum(P_D_data')/NoMonteCarlo;
```

The results of this simulation are shown in Table 3.8.

A1.8 Bootstrap model selection

Example 4.3.2. Trend estimation.

```
>> n=64; t=(0:n-1)';
>> theta=[0 0 0.035 -0.0005];
>> signal=polyval(fliplr(theta),t);
>> m=2; p=4;
```

```
>> IND=zeros(3,2^p);
>> MC=100; B=100;
>> for ii=1:MC,
       noise=randn(n,1); % Gaussian noise
       % or alternatively: noise=trnd(3,n,1);
       x=signal+noise;
       [G,AIC,MDL] = model_trend(x,B,p,m);
       [Min,ind]=min(G);IND(1,ind)=IND(1,ind)+1;
       [Min,ind]=min(AIC);IND(2,ind)=IND(2,ind)+1;
       [Min,ind]=min(MDL);IND(3,ind)=IND(3,ind)+1;
>> end;
```

The model selection procedure uses function model_trend.m which is included below.†

```
function[G,AIC,MDL,HQ,AICC,mask] = model_trend(X,B,p,m)
%
% bootstrap model selection for a linear trend
%
n=length(X); t=(0:n-1)';
H=repmat(t,1,p+1).^repmat(0:p,n,1);
theta_hat=H\X;
halfmask=fliplr(de2bi(2^(p-1):2^p-1));
mask=[flipud(~halfmask);halfmask];
msize=length(mask);
r_hat=X-H*theta_hat;
r_tilde=sqrt(n/m)*(r_hat-mean(r_hat))/sqrt(1-size(mask,2)/n);
G=zeros(1,msize);   AIC=G;   MDL=G;
for ii=1:msize,
  ind=find(mask(ii,:)~=0);
  Hm=H(:,ind);
  theta_hat_m=Hm\X;
  X_hat=Hm*theta_hat_m;
  r_hat=X-X_hat;
  r=r_hat-mean(r_hat);
  r_hat1=(r_hat-mean(r_hat))/sqrt(1-length(ind)/n);
  AIC(ii)=log(mean(r_hat1.^2))+2*(length(ind))/n;
  MDL(ii)=log(mean(r_hat1.^2))+log(n)*(length(ind))/n;
  SSE=zeros(1,B);
```

† The function model_trend.m requires function de2bi.m from the MATLAB Communication Toolbox.

```
  for b=1:B,
    r_tilde_star=boot(n,r_tilde);
    X_star=X_hat+r_tilde_star;
    thetastarm=Hm\X_star;
    Xstar=Hm*thetastarm;

    SSE(b)=mean((X-Xstar).^2);
  end;
  G(ii)=mean(SSE);
end;
```

Example 4.3.3. Harmonics in noise.

```
>> SNR=20;
>> N=64; a1=0.5; a3=0.3; B=100; M=4; MC=100;
>> n=(0:N-1)';
>> signal=a1*cos(2*pi*n/N)+a3*cos(2*pi*3*n/N);
>> amp=max(abs(signal));
>> sigma=amp*10^(-SNR/20);
>> m=2;p=3;
>> IND=zeros(2,2^p);
>> for ii=1:MC,
       noise=sigma.*randn(N,1);
       x=signal+noise;
       [G,AIC,MDL] = model_harmonics(x,B,M,m);
       [Min,ind]=min(G);
       IND(1,ind)=IND(1,ind)+1;
       [Min,ind]=min(AIC);
       IND(2,ind)=IND(2,ind)+1;
       [Min,ind]=min(MDL);
       IND(3,ind)=IND(3,ind)+1;
>> end;
```

As in the case of the trend estimation, we need to devise a separate bootstrap procedure for estimating harmonics. The MATLAB code is as follows.

```
function[G,AIC,MDL] = model_harmonics(X,B,M,m)
%
N=length(X);
n=(0:N-1)';
H=cos(repmat(2*n*pi/N,1,M).*repmat(1:M,N,1));
b_hat=inv(H'*H)*H'*X;
```

```
mask=fliplr(detobi(2^(M-1):2^M-1));
msize=length(mask);
r_hat=X-H*b_hat;
r_tilde=sqrt(N/m)*(r_hat-mean(r_hat))/sqrt(1-size(mask,2)/N);
G=zeros(1,msize);  AIC=G;  MDL=G;
for ii=1:msize,
  ind=find(mask(ii,:)~=0);
  Hm=H(:,ind);
b_hat_m=inv(Hm'*Hm)*Hm'*X;
  X_hat=Hm*b_hat_m;
  r_hat=X-X_hat;
  r=r_hat-mean(r_hat);
  r_hat1=(r_hat-mean(r_hat))/sqrt(1-length(ind)/N);
  AIC(ii)=log(mean(r_hat1.^2))+2*(length(ind))/N;
  MDL(ii)=log(mean(r_hat1.^2))+log(N)*(length(ind))/N;
  SSE=zeros(1,B);
  for b=1:B,
    r_tilde_star=boot(N,r_tilde);
    X_star=X_hat+r_tilde_star;
    bstarm=inv(Hm'*Hm)*Hm'*X_star;
    Xstar=Hm*bstarm;
    SSE(b)=mean((X-Xstar).^2);
  end;
  G(ii)=mean(SSE);
end;
```

A1.9 Noise floor estimation

```
function noisefloor()
% This is an example of the estimation of the
% noise floor in Doppler radar using the trimmed mean

disp('    ')
disp('This is an example of the optimal estimation of the')
disp('noise floor in OTHR using the trimmed mean.')
disp('    ')
disp('To determine the optimal trim we use the bootstrap.')
disp('    ')
disp('First, we simulate the target scenario using an AR model.')
disp('N=128;')
```

```
disp('X=sqrt(1/2)*randn(1,N)+sqrt(1/2)*randn(1,N)*i;')
disp('A=[1 -0.8334+0.6662i -1.0244-0.4031i 0.9036-0.2758i];')
disp('Y=filter(1,A,X);')

N=128;
X=sqrt(1/2)*randn(1,N)+sqrt(1/2)*randn(1,N)*i;
%A=[1 -2.0033+0.6018i 1.0464-0.8632i -0.0290+0.2625i]; % no "target"
A=[1 -0.8334+0.6662i -1.0244-0.4031i 0.9036-0.2758i];  % with "target"
Y=filter(1,A,X);
Ixx=abs(fft(Y)).^2/N;h=2;
Cxx=fsmooth(Ixx,bartpriest(2*h+1),h);
figure;plot(-0.5:1/(N-1):0.5,10*log10(fftshift(Cxx)));hold on
maxyaxis=max(10*log10(Cxx))+3;minyaxis=min(10*log10(Cxx));
disp('     ')
disp('Then we filter the clutter concentrated around zero-Doppler')
disp('and calculate, using the bootstrap, the optimal trimmed mean')
disp('of the noise floor')

[Xwhite]=prewhite(Y);
Xwhite=Xwhite-mean(Xwhite);
disp('    ')
disp('Please wait ....')

B=100;h=2; N=length(X);step=5;
Ixx=abs(fft(X)).^2/N;
Cxx=fsmooth(Ixx,bartpriest(2*h+1),h);
epsilon=Ixx./Cxx;
epst=epsilon./mean(epsilon);
Cxx_matrix=zeros(N,B);
for ii=1:B,
    eps_star=epst(fix(N*rand(N,1)+1));
    Ixx_star=Cxx.*eps_star;
    Cxx_matrix(:,ii)=(fsmooth(Ixx_star,bartpriest(2*h+1),h))';
end;

alpha=0:step:N/4-step; %(from 0 to 25%)
beta=0:step:N/4-step;  %(from 0 to 25%)
maxA=length(alpha);
maxB=length(beta);
var_vector=zeros(maxA,maxB);
```

```
for ii=1:maxA,
  for jj=1:maxB,
      % the two sided trimmean function can be easily adapted
      % from the Matlab trimmean.m  available in the Statistical
      % Toolbox
      z=trimmeanB(Cxx_matrix,ii*200/N,jj*200/N);
      var_vector(ii,jj)=var(z);
  end;
end;

[m1,ind1]=min(var_vector);
[m2,ind2]=min(min(var_vector));
best=[ind1(ind2) ind2];
TMbest=trimmeanB(Cxx,best(1),best(2))
Cxx=fftshift(Cxx);minyaxis1=min(10*log10(Cxx));
if minyaxis1<minyaxis,
   minyaxis=minyaxis1-3;
else
   minyaxis=minyaxis-3
end;
plot(-0.5:1/(N-1):0.5,10*log10(Cxx),'k-.')
plot(-0.5:1/(N-1):0.5,10*log10(TMbest)*ones(1,N),'r--');
hold off
legend('C_{XX}','C_{NN}','TM')
xlabel('Normalised Doppler Frequency')
ylabel('Power Spectrum [dB]')
text(-0.43,20,'Target')
axis([-0.5 0.5  minyaxis maxyaxis])
```

Several local functions are needed for the above example. They are:

```
function [C] = fsmooth(I,w,m)

if m==0,
 C=I;
else,
 s = sum(w); h = length(w);
 if (h-1)/2~=m,error('h~= 2m+1!');end;
 l = length(I); C = zeros(size(I));
 for i=-m:m
 C(max(1,1-i):min(l,l-i))=C(max(1,1-i):min(l,l-i))+...
```

```
         w(i+m+1)*I(max(i+1,1):min(l,l+i)));
  end;
  C=C/s;
end;


function[w]=bartpriest(n)

if n==1, w=1;
else
 w=2*3/4/n*(1-((0:(n-1)/2)/(n-1)*2).^2);
 w=[floor((n)/2:-1:1) w]';
end;


function[X_white]=prewhite(x,n);
%  function[X_white]=prewhite(x,n);
%
%  Prewhitening Filter
%
%  x - time series - column vector
%  n - filter order [def=10]
%
m=mean(x);
x=x(:)-m;
if exist('n')~=1, n=10; end;
if n<1, error('The order of the filter must be > 0'), end;
a=th2poly(ar(x,n,'fb0'));
X_white=filter(a,1,x)+m;
```

Appendix 2

Bootstrap MATLAB Toolbox

This appendix contains some of the MATLAB† functions that were purposely written for this book.

A2.1 Bootstrap Toolbox
Contents

---

† MATLAB is a registered trademark of The MathWorks, Inc.

bootmf.m

```
function [d] = bootmf(x,s,far,B,B1)
%    d = bootmf(x,s,far);
%
%    Inputs:
%
%    x   - observations under the model X(t)=s(t)+Z(t)
%          where Z(t) is correlated interference and
%          s(t) is a known signal;
%          can be a matrix with each set of obs. in a column
%    s   - known signal [default=ones(size(x,1),1)]
%    far - detector's false alarm rate [default=0.01]
%
%    Output:
%
%    d - decision (0=signal absent, 1=signal present) for
%        d(1), regression method
%        d(2), pivot method
%        d(3), MF
%        d(4), CFAR MF

[n,MC]=size(x);
if ~(exist('s'))
  s=ones(n,1);
end
if ~(exist('far'))
  far=0.01;
end
%B=1e2; %number of bootstrap replications
%B1=25; %number of nested bootstrap replications
lsevec=s'/(s'*s); %vector for calculating LSE

%Estimate unknown parameters and signal
theta=lsevec*x;
shat=s*theta;

%Compute centred residuals
z=x-shat;
z=z-repmat(mean(z),n,1);

xboot=zeros(n,B);
thetaboot=zeros(MC,B);
Tboot=zeros(B,MC);
stdtheta=zeros(1,B);
T=zeros(1,MC);
for i=1:MC
  xboot=shat(:,i(ones(1,B1)))+mbboot(z(:,i(ones(1,B1))));
  T(i)=theta(i)./std(lsevec*xboot);
  xboot=mbboot(z(:,i(ones(1,B))));
  thetaboot(i,:)=lsevec*xboot;
```

```
  xboot=shat(:,i(ones(1,B)))+xboot;
  thetahat=lsevec*xboot;
  shatboot=s*thetahat;
  zboot=xboot-shatboot;
  zboot=zboot-repmat(mean(zboot),n,1);
  for j=1:B
    stdtheta(j)=std(lsevec*(shatboot(:,j(ones(1,B1)))+ ...
              mbboot(zboot(:,j(ones(1,B1))))));
  end
  Tboot(:,i)=((thetahat-theta(i))./stdtheta)';
end
thetaboot=thetaboot';

d=zeros(4,MC);
d(1,:)=(sum(theta(ones(B,1),:)>thetaboot)/(B+1))>(1-far);
d(2,:)=((sum(T(ones(B,1),:)>Tboot)/(B+1))>(1-far));
d(3,:)=(theta>norminv(1-far,0,1/sqrt(s'*s)));
d(4,:)=(theta./std(z)>tinv(1-far,n-1)/sqrt(s'*s));

%%%%%%%%%%%%%%%%%% LOCAL FUNCTION %%%%%%%%%%%%%%%%%%
function xboot=mbboot(x,m)
%Returns a moving block bootstrap resample
%xboot=mbboot(x,m);
%x, original samples, one sample per column
%m, block size [default=sqrt(length(x))]
%xboot, resamples for each column of x

[n,MC]=size(x);
if ~exist('m')
  m=fix(sqrt(n));
end
xboot=zeros(n,MC);
nblocks=ceil(n/m);
x1=zeros(m*nblocks,1);
startidx=ceil((n-m+1)*rand(nblocks,MC));
m1=m-1;
k=1;
for i=1:MC
  for j=1:m:n
    x1(j:j+m1)=x(startidx(k):startidx(k)+m1,i);
    k=k+1;
  end
  xboot(:,i)=x1(1:n);
end
```

bootrsp.m

```
function[out]=bootrsp(in,B)
%    out=bootrsp(in,B)
%    Bootstrap  resampling  procedure.
%      Inputs:
%          in - input data
%           B - number of bootstrap resamples (default B=1)
%      Outputs:
%        out - B bootstrap resamples of the input data
%
%    For a vector input data of size [N,1], the  resampling
%    procedure produces a matrix of size [N,B] with columns
%    being resamples of the input vector.
%    For a matrix input data of size  [N,M], the resampling
%    procedure produces a 3D matrix of  size  [N,M,B]  with
%    out(:,:,i), i = 1,...,B, being a resample of the input
%    matrix.
%
%    Example:
%
%    out=bootrsp(randn(10,1),10);

if (exist('B')~=1), B=1;   end;
if (exist('in')~=1), error('Provide input data'); end;

s=size(in);
if length(s)>2,
   error('Input data can be a vector or a 2D matrix only');
end;
if min(s)==1,
   out=in(ceil(max(s)*rand(max(s),B)));
else
   out=in(ceil(s(1)*s(2)*rand(s(1),s(2),B)));
end;
```

bootrsp2.m

```
function[out1,out2]=bootrsp2(in1,in2,B)
%       [out1,out2]=bootrsp2(in1,in2,B)
%   Bootstrap  resampling  procedure for bivariate data.
%
%     Inputs:
%        in1 - input data (first variate)
%        in2 - input data (second variate). If in2 is not
%              provided the function runs bootrsp.m by default.
%          B - number of bootstrap resamples (default B=1)
%     Outputs
%        out1 - B bootstrap resamples of the first variate
%        out2 - B bootstrap resamples of the second variate
%
%   For a vector input data of size [N,1], the  resampling
%   procedure produces a matrix of size [N,B] with columns
%   being resamples of the input vector.
%
%   Example:
%
%   [out1,out2]=bootrsp2(randn(10,1),randn(10,1),10);

if (exist('B')~=1), B=1;  end;
if (exist('in2')~=1) & (exist('in1')==1),
   out1=bootrsp(in1,B); out2=0;
   return
end;
if (exist('in1')~=1), error('Provide input data'); end;
s1=size(in1); s2=size(in2);

if length(s1)>2 | length(s2)>2,
   error('Input data can be vectors or 2D matrices only');
end;

if any(s1-s2)~=0 & any(s1-fliplr(s2))~=0,
   error('Input vectors or matrices must be of the same size')
end;
if s1==fliplr(s2),
   in2=in2.';
end;
if min(s1)==1,
   ind=ceil(max(s1)*rand(max(s1),B));
   out1=in1(ind); out2=in2(ind);
else
   ind=ceil(s1(1)*s1(2)*rand(s1(1),s1(2),B));
   out1=in1(ind);
   out2=in2(ind);
end;
```

boottest.m†

```
function[H]=boottest(x,statfun,vzero,type,alpha,B1,B2,varargin)
%
%       D=boottest(x,statfun,v_0,type,alpha,B1,B2,PAR1,...)
%
%       Hypothesis test for a characteristic (parameter) 'v'
%       of an unknown distribution  based on the bootstrap
%       resampling procedure and pivoted test statistics
%
%       Inputs:
%               x - input vector data
%         statfun - the estimator of the parameter given as a Matlab function
%           v_0   - the value of 'v' under the null hypothesis
%           type  - the type of hypothesis test.
%
%                   For type=1:   H: v=v_0    against K: v~=v_0
%                   (two-sided hypothesis test)
%                   For type=2:   H: v<=v_0   against K: v>v_0
%                   (one-sided hypothesis test)
%                   For type=3:   H: v>=v_0   against K: v<v_0
%                   (one-sided hypothesis test)
%                   (default type=1)
%         alpha   - the level of the test (default alpha=0.05)
%             B1 - number of bootstrap resamplings
%                  (default B1=99)
%             B2 - number of bootstrap resamplings for variance
%                  estimation (nested bootstrap)
%                  (default B2=25)
%       PAR1,... - other parameters than x to be passed to statfun
%
%       Outputs:
%             D - The output of the test.
%                   D=0: retain the null hypothesis
%                   D=1: reject the null hypothesis
%
%       Example:
%
%       D = boottest(randn(10,1),'mean',0);

pstring=varargin;
if (exist('B2')~=1), B2=25; end;
if (exist('B1')~=1), B1=99; end;
if (exist('alpha')~=1), alpha=0.05; end;
if (exist('type')~=1), type=1; end;
if (exist('vzero')~=1),
  error('Provide the value of the parameter under the null hypothesis');
end;
```

---

† This function requires **bootstrp.m** from the MATLAB Statistical Toolbox

```
x=x(:);
vhat=feval(statfun,x,pstring{:});
bstat=bootstrp(B2,statfun,x,pstring{:});

if type==1,
  T=abs(vhat-vzero)./std(bstat);
else
  T=(vhat-vzero)./std(bstat);
end;

[vhatstar,ind]=bootstrp(B1,statfun,x,pstring{:});
bstats=bootstrp(B2,statfun,x(ind),pstring{:});

M=ceil((B1+1)*(1-alpha));

if type==1,
  tvec=abs(vhatstar-vhat)./std(bstats)';
  st=sort(tvec);
  if T>st(M), H=1; else H=0; end;
elseif type==2,
  tvec=(vhatstar-vhat)./std(bstats)';
  st=sort(tvec);
  if T>st(M), H=1; else H=0; end;
elseif type==3,
  tvec=(vhatstar-vhat)./std(bstats)';
  st=sort(tvec);
  if T<st(M), H=1; else H=0; end;
end;
```

boottestnp.m†

```
function[H]=boottestnp(x,statfun,vzero,type,alpha,B,varargin)
%
%       D=boottestnp(x,statfun,v_0,type,alpha,B,PAR1,...)
%
%       Hypothesis test for a characteristic (parameter) 'v'
%       of an unknown distribution  based on the bootstrap
%       resampling procedure and unpivoted test statistics
%
%     Inputs:
%            x - input vector data
%       statfun - the estimator of the parameter given as a Matlab function
%         v_0  - the value of vartheta under the null hypothesis
%        type - the type of hypothesis test.
%
%               For type=1:   H: v=v_0    against K: v~=v_0
%                (two-sided hypothesis test)
%               For type=2:   H: v<=v_0   against K: v>v_0
%                (one-sided hypothesis test)
%               For type=3:   H: v>=v_0   against K: v<v_0
%                (one-sided hypothesis test)
%                (default type=1)
%       alpha  - determines the level of the test
%                (default alpha=0.05)
%            B - number of bootstrap resamplings
%                (default B1=99)
%     PAR1,... - other parameters than x to be passed to statfun
%
%     Outputs:
%            D - The output of the test.
%                D=0: retain the null hypothesis
%                D=1: reject the null hypothesis
%
%     Example:
%
%       D = boottestnp(randn(10,1),'mean',0);

pstring=varargin;
if (exist('B')~=1), B=99; end;
if (exist('alpha')~=1), alpha=0.05; end;
if (exist('type')~=1), type=1; end;
if (exist('vzero')~=1),
  error('Provide the value of the parameter under the null hypothesis');
end;

x=x(:);
vhat=feval(statfun,x,pstring{:});
```

† This function requires **bootstrp.m** from the MATLAB Statistical Toolbox

```
if type==1,
  T=abs(vhat-vzero);
else
  T=vhat-vzero;
end;

[vhatstar,ind]=bootstrp(B,statfun,x,pstring{:});

M=floor((B+1)*(1-alpha));

if type==1,
  tvec=abs(vhatstar-vhat);
  st=sort(tvec);
  if T>st(M), H=1; else H=0; end;
elseif type==2,
  tvec=vhatstar-vhat;
  st=sort(tvec);
  if T>st(M), H=1; else H=0; end;
elseif type==3,
  tvec=vhatstar-vhat;
  st=sort(tvec);
  if T<st(M), H=1; else H=0; end;
end;
```

boottestvs.m†

```
function[H]=boottestvs(x,statfun,vzero,type,alpha,B1,B2,B3,varargin)
%       D=boottestvs(x,statfun,v_0,type,alpha,B1,B2,B3,PAR1,...)
%
%
%        Hypothesis test for a characteristic (parameter) 'v'
%        of an unknown distribution  based on the bootstrap
%        resampling procedure and variance stabilisation (VS).
%
%      Inputs:
%            x - input vector data
%       statfun - the estimator of the parameter given as a Matlab function
%          v_0  - the value of vartheta under the null hypothesis
%          type - the type of hypothesis test.
%
%                 For type=1:   H: v=v_0   against K: v~=v_0
%                 (two-sided hypothesis test)
%                 For type=2:   H: v<=v_0  against K: v>v_0
%                 (one-sided hypothesis test)
%                 For type=3:   H: v>=v_0  against K: v<v_0
%                 (one-sided hypothesis test)
%                 (default type=1)
%        alpha  - determines the level of the test
%                 (default alpha=0.05)
%          B1   - number of bootstrap resamplings for VS
%                 (default B1=100)
%          B2   - number of bootstrap resamplings for VS
%                 (default B2=25)
%          B3   - number of bootstrap resamplings
%                 (default B3=99)
%     PAR1,... - other parameters than x to be passed to statfun
%
%      Outputs:
%            D - The output of the test.
%                D=0: retain the null hypothesis
%                D=1: reject the null hypothesis
%
%      Example:
%
%      D = boottestvs(randn(10,1),'mean',0);

pstring=varargin;
if (exist('B3')~=1), B3=99; end;
if (exist('B2')~=1), B2=25; end;
if (exist('B1')~=1), B1=100; end;
if (exist('alpha')~=1), alpha=0.05; end;
if (exist('type')~=1), type=1; end;
if (exist('vzero')~=1),
   error('Provide the value of the parameter under the null hypothesis');
```

† This function requires **bootstrp.m** from the MATLAB Statistical Toolbox

```
end;

x=x(:);
vhat=feval(statfun,x,pstring{:});

[vhatstar,ind]=bootstrp(B1,statfun,x,pstring{:});
bstats=bootstrp(B2,statfun,x(ind),pstring{:});
sigmastar2=var(bstats);

[statsort,sigmasort,sigmasm2]=smooth(vhatstar',sigmastar2,B1/200);

a=statsort;
b=sigmasm2.^(-1/2);
h=zeros(1,B1);
h(1)=0;
for i=2:B1,
    h(i)=h(i-1)+(a(i)-a(i-1))*(b(i)+b(i-1))/2;
end;

[vhatstar1,ind1]=bootstrp(B3,statfun,x,pstring{:});

ind=find(vhatstar1>=a(1) & vhatstar1<=a(B1));
ind1=find(vhatstar1<a(1));
ind2=find(vhatstar1>a(B1));
newv=vhatstar1(ind);
newvs=vhatstar1(ind1);
newvl=vhatstar1(ind2);
hvec(ind)=interp1(a,h,newv)';
hvec(ind1)=(h(2)-h(1))/(a(2)-a(1))*(newvs-a(1))+h(1);
hvec(ind2)=(h(B1)-h(B1-1))/(a(B1)-a(B1-1))*(newvl-a(B1-1))+h(B1-1);

p=find(a>vhat);
if isempty(p)
  hvhat=(h(B1)-h(B1-1))/(a(B1)-a(B1-1))*(vhat-a(B1-1))+h(B1-1);
elseif p(1)==1,
  hvhat=(h(2)-h(1))/(a(2)-a(1))*(vhat-a(1))+h(1);
else
  hvhat=interp1(a,h,vhat);
end;

p=find(a>vzero);
if isempty(p)
  hvzero=(h(B1)-h(B1-1))/(a(B1)-a(B1-1))*(vzero-a(B1-1))+h(B1-1);
elseif p(1)==1,
  hvzero=(h(2)-h(1))/(a(2)-a(1))*(vzero-a(1))+h(1);
else
  hvzero=interp1(a,h,vzero);
end;
M=floor((B3+1)*(1-alpha));
if type==1,
```

```
    Tstar=abs(hvec-hvhat);
    T=abs(hvhat-hvzero);
    ST=sort(Tstar);
    if T>ST(M), H=1; else H=0; end;
elseif type==2,
    Tstar=(hvec-hvhat);
    T=(hvhat-hvzero);
    ST=sort(Tstar);
    if T>ST(M), H=1; else H=0; end;
elseif type==3,
    Tstar=(hvec-hvhat);
    T=(hvhat-hvzero);
    ST=sort(Tstar);
    if T<ST(M), H=1; else H=0; end;
end;
```

bpestcir.m

```
function[est,estvar]=bpestcir(x,estfun,L1,M1,Q1,L2,M2,Q2,B,varargin)
%         [est,estvar]=bpestcir(X,estfun,L1,M1,Q1,L2,M2,Q2,B,PAR1,...)
%
%         The program calculates the estimate and the variance
%         of an estimator of a parameter from the input vector X.
%         The algorithm is based on a circular block bootstrap
%         and is suitable when the data are weakly correlated.
%
%     Inputs:
%             x - input vector data
%        estfun - the estimator of the parameter given as a Matlab function
%            L1 - number of elements in the first block (see "segmcirc.m")
%            M1 - shift size in the first block
%            Q1 - number of segments in the first block
%            L2 - number of elements in the second block (see "segmcirc.m")
%            M2 - shift size in the second block
%            Q2 - number of segments in the second block
%             B - number of bootstrap resamplings (default B=99)
%    PAR1,... - other parameters than x to be passed to estfun
%
%     Outputs:
%           est - estimate of the parameter
%        estvar - variance of the estimator
%
%     Example:
%
%        [est,estvar]=bpestcir(randn(1000,1),'mean',50,50,50,10,10,10);

pstring=varargin;
if (exist('B')~=1), B=99; end;
x=x(:);
QL=segmcirc(x,L1,M1,Q1);
estm=feval(estfun,QL,pstring{:});
beta=segmcirc(estm,L2,M2,Q2);
ind=bootrsp(1:Q2,B);
Y=beta(ind);
estsm=mean(Y);
est=mean(estsm);
estvar=var(estsm);
```

bpestdb.m

```
function[est,estvar]=bpestdb(x,estfun,L1,M1,L2,M2,B,varargin)
%          [est,estvar]=bpestdb(X,estfun,L1,M1,L2,M2,B,PAR1,...)
%
%         The program calculates the estimate and the variance
%         of an estimator of a parameter from the input vector X.
%         The algorithm is based on a double block bootstrap
%         and is suitable when the data are weakly correlated.
%
%      Inputs:
%              x - input vector data
%         estfun - the estimator of the parameter given as a Matlab function
%             L1 - number of elements in the first block (see "segments.m")
%             M1 - shift size in the first block
%             L2 - number of elements in the second block (see "segments.m")
%             M2 - shift size in the second block
%              B - number of bootstrap resamplings (default B=99)
%      PAR1,... - other parameters than x to be passed to estfun
%
%      Outputs:
%            est - estimate of the parameter
%         estvar - variance of the estimator
%
%      Example:
%
%      [est,estvar]=bpestdb(randn(1000,1),'mean',50,50,10,10);

pstring=varargin;
if (exist('B')~=1), B=99; end;
x=x(:);
[QL,Q]=segments(x,L1,M1);
estm=feval(estfun,QL,pstring{:});
[beta,q]=segments(estm,L2,M2);
ind=bootrsp(1:q,B);
Y=beta(ind);
estsm=mean(Y);
est=mean(estsm);
estvar=var(estsm);
```

bspecest.m

```
function[Cxx_hat,f_low,f_up]=bspecest(X,h,kerneltype,alpha,B)
%    function[Cxx_hat,f_low,f_up]=bspecest(X,h,kerneltype,alpha,B)
%
%  The procedure for bootstrapping kernel spectral density
%  estimates based on resampling from the periodogram of
%  the original data.
%             X          - input data
%             h          - spectrum estimation bandwidth
%             kerneltype - type of kernel
%             alpha      - 100(1-alpha)% confidence interval
%             B          - number of bootstrap resamplings
%             Cxx_hat    - spectral estimate
%             f_low      - lower confidence band
%             f_up       - upper confidence band

T=length(X);
Ixx=abs(fft(X)).^2/T/2/pi;
m=round((T*h-1)/2);
K=window(kerneltype,2*m+1);
K=K/sum(K);
Cxx_hat=ksde(Ixx,K);
epsilon=Ixx(1:floor(T/2))./Cxx_hat(1:floor(T/2));
epst=epsilon/mean(epsilon);
eps_star=bootrsp(epst,B);
Cxx_hat_star=zeros(T,B);
for b=1:B,
   Ixx_star=Cxx_hat.*[eps_star(:,b); flipud(eps_star(:,b))];
   Cxx_hat_star(:,b)=ksde(Ixx_star,K);
end;
baps=(Cxx_hat_star-Cxx_hat*ones(1,B))./(Cxx_hat*ones(1,B));
baps=sort(baps,2);
lo=floor(B*(1-alpha)/2);
up=B-lo+1;
f_lo=1./(1+baps(:,up)).*Cxx_hat;
f_up=1./(1+baps(:,lo)).*Cxx_hat;
%%%%%%%%%%%%%%%%%%%%%%%%% LOCAL FUNCTIONS %%%%%%%%%%%%%%%%%%%
function [Cxx]=ksde(Ixx,K,zm)
%KSDE  Computes kernel spectral density estimate
%
%  Inputs
%    Ixx : periodogram
%      K : kernel
%     zm : if zm=1, Cxx(1)=0
%
%  Outputs
%    Cxx : kernel spectral density estimate
%
%  Assumptions
%    1. Ixx(1) = 0
```

```
%     2. 2M < T, where T is the number of observations
%     3. length(K) = 2M+1 is odd
%     4. sum(K) ~= 0

% If 2M >= T, then Ixx has to be concatenated to itself until
% it is longer than 2M, after which K must be zero padded to
% this new length T' = T*floor((2M+1)/T)

if ~exist('zm'), zm=0; end;

column=size(Ixx,1)>1;
Ixx=Ixx(:); K=K(:);

M=(length(K)-1)/2;
T=length(Ixx);

if zm, Ixx(1)=0; end;
if even(T), Ixx(T/2+1)=0; end; % special case
K(end+1:T)=0; % Pad to length T
% periodically convolve
Cxx=real(ifft(fft(Ixx).*fft(K)));
% adjust so that Cxx(1) corresponds to zero frequency
Cxx=[Cxx(M+1:end); Cxx(1:M)];
if zm, Cxx(1)=0; end; % return Cxx(1) to 0
% if T is even, Cxx(T/2+1) is the windowed average
% around T/2+1 ignoring T/2+1

% determine the correct scaling constant
mask=ones(T,1);
if zm, mask(1)=0; end;
if even(T), mask(T/2+1)=0; end;
scale=real(ifft(fft(mask).*fft(K)));
scale=[scale(M+1:end); scale(1:M)];

% scale by sum of window function
Cxx=Cxx./scale;

if ~column, Cxx=Cxx.'; end;
%%%%%%%%%%%%%%%%%%%%%%%%%%%%%%%%%%%%%%%%%%%%%%%%%%%%%%%%%%%%%%
function [e]=even(x)

e = (rem(x,2)==0).*(x>0);
```

bspecest2.m

```
function[TN,f_low,f_up]=bspecest2(X,M,L,b,h,alpha,k,B)
%          function[TN,f_low,f_up]=bspecest2(X,M,L,b,h,alpha,k,B)
%
%          The program calculates the bootstrap confidence Bands
%          for Spectra of stationary weakly dependent time series
%          (input vector X). The algorithm is based on a double
%          block bootstrap.
%
%             X     - input vector
%             M     - number of elements in segments in the first block
%             L     - number of overlapping elements in the first block
%             b     - number of elements in segments in the second block
%             h     - number of overlapping elements in the second block
%             alpha - 100(1-alpha)% confidence intervals
%             k     - number of bootstrap samples
%             B     - number of the Monte Carlo evaluations
%

  P_star=zeros(b,k);
  T_hat_star =zeros(M,B);
  pr_star=zeros(M,B);
  [s1,s2]=size(X);
  if s1~=1,
    error('Input must be a vector')
  end;
  [BiML,Q]=segments(X,2*M,2*L);
  TiML=abs(fft(BiML)).^2/M/2;
  TN=mean(TiML(1:M,:)')';
  for w=1:M,
    [BeJ,q]=segments(TiML(w,:),b,h);
    for j=1:B,
      ind=boot(k,1:q);
      P_star(1:b,:)=BeJ(1:b,ind);
      T_star=reshape(P_star,1,k*b);
      T_hat_star(w,j)=mean(T_star);
    end;
  end;
  for i=1:B,
    pr_star(:,i)=(T_hat_star(:,i)-TN)./TN;
  end;
  Pr_sort=sort(pr_star')';
  qlo=fix(B*alpha/2);
  qup=B-qlo+1;
  cL=Pr_sort(:,qlo);
  cU=Pr_sort(:,qup);
  f_low=(1+cL).*TN;
  f_up=(1+cU).*TN;
```

confint.m†

```
function[Lo,Up]=confint(x,statfun,alpha,B1,B2,varargin)
%
%       [Lo,Up]=confint(x,statfun,alpha,B1,B2,PAR1,...)
%
%       Confidence interval of the estimator of a parameter
%       based on the bootstrap percentile-t method
%
%     Inputs:
%             x - input vector data
%       statfun - the estimator of the parameter given as a Matlab function
%       alpha  - level of significance (default alpha=0.05)
%          B1 - number of bootstrap resamplings (default B1=199)
%          B2 - number of bootstrap resamplings for variance
%               estimation (nested bootstrap) (default B2=25)
%     PAR1,... - other parameters than x to be passed to statfun
%
%     Outputs:
%          Lo - The lower bound
%          Up - The upper bound
%
%     Example:
%
%       [Lo,Up] = confint(randn(100,1),'mean');

pstring=varargin;
if (exist('B2')~=1), B2=25; end;
if (exist('B1')~=1), B1=199; end;
if (exist('alpha')~=1), alpha=0.05; end;

x=x(:);
vhat=feval(statfun,x,pstring{:});
[vhatstar,ind]=bootstrp(B1,statfun,x,pstring{:});

if length(pstring)~=0,
  if length(pstring{:})==length(x)
     newpstring=pstring{:};
     bstats=bootstrp(B2,statfun,x(ind),newpstring(ind));
  else
     bstats=bootstrp(B2,statfun,x(ind),pstring{:});
  end;
else
  bstats=bootstrp(B2,statfun,x(ind),pstring{:});
end;
bstat=bootstrp(B2,statfun,x,pstring{:});
sigma1=std(bstat);

q1=floor(B1*alpha*0.5);
```

† This function requires **bootstrp.m** from the MATLAB Statistical Toolbox

```
q2=B1-q1+1;
sigma=std(bstats)';
tvec=(vhatstar-vhat)./sigma;
[st,ind]=sort(tvec);
lo=st(q1);
up=st(q2);
Lo=vhat-up*sigma1;
Up=vhat-lo*sigma1;
```

confintp.m†

```
function[Lo,Up]=confintp(x,statfun,alpha,B1,varargin)
%
%        [Lo,Up]=confintp(x,statfun,alpha,B1,PAR1,...)
%
%        Confidence interval of the estimator of a parameter
%        based on the bootstrap percentile method
%
%     Inputs:
%           x - input vector data
%     statfun - the estimator of the parameter given as a Matlab function
%       alpha - level of significance (default alpha=0.05)
%          B1 - number of bootstrap resamplings (default B1=199)
%     PAR1,... - other parameters than x to be passed to statfun
%
%     Outputs:
%          Lo - The lower bound
%          Up - The upper bound
%
%     Example:
%
%        [Lo,Up] = confintp(randn(100,1),'mean');

pstring=varargin;
if (exist('B1')~=1), B1=199; end;
if (exist('alpha')~=1), alpha=0.05; end;

x=x(:);
vhat=feval(statfun,x,pstring{:});
vhatstar=bootstrp(B1,statfun,x,pstring{:});

q1=floor(B1*alpha*0.5);
q2=B1-q1+1;
st=sort(vhatstar);
Lo=st(q1);
Up=st(q2);
```

† This function requires **bootstrp.m** from the MATLAB Statistical Toolbox

confinth.m†

```
function[Lo,Up]=confinh(x,statfun,alpha,B1,varargin)
%
%       [Lo,Up]=confinh(x,statfun,alpha,B1,PAR1,...)
%
%       Confidence interval of the estimator of a parameter
%       based on the bootstrap hybrid method
%
%     Inputs:
%            x - input vector data
%      statfun - the estimator of the parameter given as a Matlab function
%       alpha  - level of significance (default alpha=0.05)
%           B1 - number of bootstrap resamplings (default B1=199)
%      PAR1,... - other parameters than x to be passed to statfun
%
%     Outputs:
%           Lo - The lower bound
%           Up - The upper bound
%
%     Example:
%
%     [Lo,Up] = confinth(randn(100,1),'mean');

pstring=varargin;

if (exist('B1')~=1), B1=199; end;
if (exist('alpha')~=1), alpha=0.05; end;

x=x(:);
vhat=feval(statfun,x,pstring{:});
[vhatstar,ind]=bootstrp(B1,statfun,x,pstring{:});

q1=floor(B1*alpha*0.5);
q2=B1-q1+1;
tvec=vhatstar-vhat;
st=sort(tvec);
Lo=vhat-st(q2);
Up=vhat-st(q1);
```

---

† This function requires **bootstrp.m** from the MATLAB Statistical Toolbox

jackest.m

```
function[est,estall]=jackest(x,estfun,h,varargin)
%        [est,estall]=jackest(x,estfun,h,PAR1,...)
%
%        Parameter estimation based on the "Jackknife" procedure
%
%      Inputs:
%             x - input vector data
%        estfun - the estimator of the parameter given as a Matlab function
%             h - number of elements in a block that is to be deleted
%                 see jackrsp.m (defult h=1)
%     PAR1,... - other parameters than x to be passed to estfun
%
%      Outputs:
%           est - the jackknifed estimate
%        estall - the estimate based on the whole sample
%
%      Example:
%
%      [est,estall]=jackest(randn(10,1),'trimmean',1,20);

pstring=varargin;
if (exist('h')~=1), h=1; end;
x=x(:);
N=length(x);
estall=feval(estfun,x,pstring{:});
esti=feval(estfun,jackrsp(x,h),pstring{:});
%keyboard
psv=N*estall-(N-1).*esti;
est=mean(psv);
```

jackrsp.m

```
function[y]=jackrsp(x,h)
%       y=jackrsp(x,h)
%
%       The procedure known as a "Jackknife" forms a matrix of size
%       (g-1)*h by g from the input vector x of length g*h. The
%       input vector is first divided into "g" blocks of size "h".
%       Each column of the matrix is formed by deleting a block
%       from the input. The standard version of the Jackknife is
%       when h=1.
%
%       Inputs:
%           x - input vector data
%           h - number of elements in a block that is to be deleted
%               (default h=1)
%       Output:
%           y - the output matrix of the data
%
%       Example:
%
%       y=jackrsp(randn(10,1));

if (exist('h')~=1), h=1; end;

x=x(:);
N=length(x);
g=N/h;
if rem(N,h)~=0,
  error('The length of the input vector must be divisible by h')
  return
end;
y=zeros((g-1)*h,g);
for ii=1:g,
  y(:,ii)=x([1:ii*h-h ii*h+1:N]);
end;
```

segmcirc.m

```
function[y]=segmcirc(x,L,M,Q)
%        y=segmcirc(X,L,M,Q)
%
%        Given the data samples X=(x_1,x_2,...,x_N),
%        the program obtains Q overlapping (M<L) or
%        non-overlapping (M>=L) segments, each of L samples
%        in the form of a matrix "y" of L rows and Q columns.
%        The data X_i is "wrapped" around in a circle, that is,
%        define (for i>N) X_i=X_iN, where iN=i(mod N).
%
%            ----------------
%        .....|_____L_____|  .....
%        .....|____M____|_____L_____|  .....
%        .....|___ M ___|___ M ___|_____ L _____|  .....
%
%        The procedure is used for the circular block bootstrap.
%
%        Inputs:
%              X - input vector data
%              L - number of elements in a segment
%              M - shift size (i.e. L-M is the size of overlap)
%              Q - number of desired segments
%        Output:
%              y - the output matrix of the data

x=x(:);
N=length(x);
y=zeros(L,Q);
Ny=Q*M+L-1;
Y=zeros(Ny,1);
r=0;
for ii=1:Ny,
  Y(ii)=x(ii-rem(N*r,ii));
  if ii/N==r+1,
      r=r+1;
  end;
end;
for ii=1:Q,
  y(:,ii)=Y((ii-1)*M+1:(ii-1)*M+L);
end;
```

segments.m

```
function[y,q]=segments(x,L,M)
%       [y,Q]=segments(X,L,M)
%
%       Given the data samples X=(x_1,x_2,...,x_N),
%       the program obtains Q overlapping (M<L) or
%       non-overlapping (M>=L) segments, each of L samples
%       in the form of a matrix "y" of L rows and Q columns.
%
%          ----------------
%         |_____ L _____| .....
%         |___ M ___|_____ L _____| .....
%         |___ M ___|___ M ___|_____ L _____| .....
%
%       The procedure is used for the block of blocks bootstrap.
%
%       Inputs:
%           X - input vector data
%           L - number of elements in a segment
%           M - shift size (i.e. L-M is the size of overlap)
%       Output:
%           y - the output matrix of the data
%           Q - number of output segments

x=x(:);
N=length(x);
q=fix((N-L)/M)+1;
y=zeros(L,q);
for ii=1:q,
  y(:,ii)=x((ii-1)*M+1:(ii-1)*M+L);
end;
```

smooth.m

```
function [x_sort, y_sort, y_sm] = smooth(x, y, w)
%              [x_sort, y_sort, y_sm] = smooth(x, y, w);
%              A running line smoother that fits the data by linear
%              least squares. Used to compute the variance stabilising
%              transformation.
%
%    Inputs:
%              x - one or more columns of covariates
%              y - one column of response for each column of covariate
%              w - span, proportion of data in symmetric centred window
%
%    Outputs:
%       x_sort - sorted columns of x
%       y_sort - values of y associated with x_sort
%         y_sm - smoothed version of y
%
%  Note: If inputs are row vectors, operation is carried out row-wise.

if any(size(x) ~= size(y))
    error('Input matrices must be the same length.'),
end

[nr,nc] = size(x);
n=nr;
if (nr==1) x=x';y=y';n=nc;nc=1; end
y_sm = zeros(n,nc);
[x_sort,order] = sort(x);
for i = 1:nc y_sort(:,i) = y(order(:,i),i); end
k = fix(w*n/2);

for i = 1:n
    window = max(i-k,1):min(i+k,n);
    xwin = x_sort(window,:);
    ywin = y_sort(window,:);
    xbar = mean(xwin);
    ybar = mean(ywin);
    copy = ones(length(window),1);
    x_mc = xwin - copy*xbar;     % mc = mean-corrected
    y_mc = ywin - copy*ybar;
    y_sm(i,:) = sum(x_mc.*y_mc)./sum(x_mc.*x_mc) .* (x_sort(i,:)-xbar) + ybar;
end

if (nr==1)
    x_sort=x_sort';y_sort=y_sort';y_sm=y_sm';
end
```

References

Abutaleb, A. S. (2002). Number Theory and Bootstrapping for Phase Unwrapping. *IEEE Transactions on Circuits and Systems I: Fundamental Theory and Applications*, 49(5): 632–638.

Akaike, H. (1970). Statistical Predictor Identification. *Annals of the Institute of Statistical Mathematics*, 22: 203–217.

(1974). A New Look at the Statistical Model Identification. *IEEE Transactions on Automatic Control*, 19: 716–723.

(1977). An Objective Use of Bayesian Models. *Annals of the Institute of Statistical Mathematics*, 29: 9–20.

(1978). Comments on "On Model Structure Testing in System Identification". *International Journal of Control*, 27: 323–324.

Anderson, R. H. and Krolik, J. L. (1998a). Over-the-Horizon Radar Target Localization Using a Hidden Markov Model Estimated from Ionosonde Data. *Radio-Science*, 33(4): 1199–1213.

(1998b). The Performance of Maximum Likelihood Over-The-Horizon Radar Coordinate Registration. In *Proceedings of the IEEE International Conference on Acoustics, Speech and Signal Processing, ICASSP-98*, 4: 2481–2484, Seattle, USA.

(1999). Multipath Track Association for Over-The-Horizon Radar Using a Bootstrapped Statistical Ionospheric Model. In *Proceedings of 33rd Asilomar Conference on Signals, Systems and Computers*, 1: 8–14, Pacific Grove, California.

Anderson, S. J. (1986). Remote Sensing with JINDALEE Skywave Radar. *IEEE Journal of Oceanic Engineering*, 11(2): 158–163.

Anderson, T. W. (1963). Asymptotic Theory for Principal Component Analysis. *Annals of Mathematical Statistics*, 34: 122–148.

(1984). *An Introduction to Multivariate Statistical Analysis*. John Wiley and Sons, New York.

Andrews, D. W. K. (2002). Higher-order Improvements of a Computationally Attractive k-step Bootstrap for Extremum Estimators. *Econometrica*, 70(1): 119–162.

Archer, G. and Chan, K. (1996). Bootstrapping Uncertainty in Image Analysis. In *Proceedings in Computational Statistics, Physica-Verlag*, 193–198, Heidelberg, Germany.

Arnholt, A. T., Johnson, D. B., and Hebert, J. L. (1998). Improving Estimators by Bootstrapping a Function of the Coefficient of Variation. *ASA Proceedings of the Statistical Computing Section*, 54–56.

Aronsson, M., Arvastson, L., Holst, J., Lindoff, B., and Svensson, A. (1999). Bootstrap Control. In *Proceedings of the 1999 American Control Conference*, 6: 4516–4521.

Athreya, K. B. (1987). Bootstrap of the Mean in the Infinite Variance Case. *The Annals of Statistics*, 15(2): 724–731.

Babu, G. J. and Singh, K. (1983). Inference on Means using the Bootstrap. *The Annals of Statistics*, 11: 999–1003.

Banga, C. and Ghorbel, F. (1993). Optimal Bootstrap Sampling for Fast Image Segmentation: Application to Retina Image. In *Proceedings of the IEEE International Conference on Acoustics, Speech and Signal Processing, ICASSP-93*, 5: 638–641, Minnesota.

Barbe, P. and Bertail, P. (1995). *The Weighted Bootstrap*, Volume 98 of *Lecture notes in Statistics*. Springer, New York.

Barkat, B., Zoubir, A. M., and Brown, C. L. (2000). Application of Time-Frequency Techniques for the Detection of Anti-Personnel Landmines. In *Proceedings of the Tenth IEEE Workshop on Statistical Signal and Array Processing*, 594–597.

Barrick, D. E. and Snider, J. B. (1977). The Statistics of HF Sea-Echo Doppler Spectra. *IEEE Transactions on Antennas and Propagation*, 25(1): 19–28.

Bello, M. G. (1998). Comparison of Parametric and Nonparametric ROC Confidence Bound Construction in the Context of Acoustic/Magnetic Fusion Systems for Mine-Hunting. In Dubey, A. C., Harvey, J. F., and Broach, J. T., Editors, *Detection and Remediation Technologies for Mines and Minelike Targets III*, 3392: 1162–1177, San Diego. Proceedings of SPIE.

Beran, R. (1986). Simulated power functions. *Annals of Statistics*, 14(1): 151–173.

Beran, R. and Srivastava, M. (1985). Bootstrap Tests and Confidence Regions for Functions of a Covariance Matrix. *The Annals of Statistics*, 13(1): 95–115.

(1987). Correction: Bootstrap tests and confidence regions for functions of a covariance matrix. *The Annals of Statistics*, 15(1): 470–471.

Besson, V. and Shenton, A. T. (2000). Interactive Parameter Space Design for Robust Performance of MISO Control Systems. *IEEE Transactions on Automatic Control*, 45(10): 1917–1924.

Bhattacharya, R. N. and Rao, R. R. (1976). *Normal Approximation and Asymptotic Expansion*. Wiley, New York.

Bhide, V. M., Pivoso, M. J., and Kosanovich, K. A. (1995). Statistics on Reliability of Neural Networks Estimates. In *Proceedings of the IEEE 1995 American Control Conference*, 3: 1877–1881, Seattle, WA, USA.

Bickel, P. J. and Freedman, D. A. (1981). Some Asymptotic Theory for the Bootstrap. *The Annal of Statistics*, 9: 1196–1217.

Bickel, P. J. and Yahav, J. A. (1988). Richardson Extrapolation and the Bootstrap. *Journal of the American Statistical Association*, 83: 387–393.

Boashash, B. (1992a). Interpreting and Estimating the Instantaneous Frequency of a Signal – Part I: Fundamentals. *Proceedings of the IEEE*, 519–538, April 1992.

(1992b). Interpreting and Estimating the Instantaneous Frequency of a Signal – Part II: Algorithms. *Proceedings of the IEEE*, 539–569, April 1992.

Boashash, B., O'Shea, P. J., and Ristic, B. (1991). A Stat/Computational Comparison of Some Algorithms for Instantaneous Frequency Estimation

Proceedings of the IEEE Conference on Acoustics, Speech and Signal Processing, 5: 3193–3196, Toronto, Canada.

Böhme, J. F. (1991). Array Processing In *Advances in Spectrum Estimation*, Haykin, S., Editor, 2: 1–63, Prentice Hall, Englewood Cliffs.

Böhme, J. F. and Maiwald, D. (1994). Multiple Wideband Signal Detection and Tracking from Towed Array Data. In Blanke, M. and Söderström, T., editors, *SYSID 94, 10th IFAC Symposium on System Identification*, 1: 107–112, Copenhagen.

Booth, J. G., Hall, P., and Wood, A. T. A. (1993). Balanced Importance Resampling for the Bootstrap. *The Annals of Statistics*, 21: 286–298.

Bose, A. (1988). Edgeworth Correction by Bootstrap in Autoregressions. *The Annals of Statistics*, 16: 1709–1722.

Bratley, P., Fox, B. L., and Schrage, E. L. (1983). *A Guide to Simulation*. Springer-Verlag, New York.

Brcich, R. F., Zoubir, A. M., and Pelin, P. (2002). Detection of Sources using Bootstrap Techniques *IEEE Transactions on Signal Processing*, 50(2): 206–215.

Brillinger, D. R. (1981). *Time Series: Data Analysis and Theory*. Holden-Day, San Francisco.

(1983). The Finite Fourier Transform of a Stationary Process. In *Handbook of Statistics III: Time series in the frequency domain*, Brillinger, D. R. and Krishnaiah, P. R., Editors, 237–278, North-Holland, Amsterdam, New York.

Bühlmann, P. (1994). Blockwise Bootstrapped Empirical Process for Stationary Sequences. *The Annals of Statistics*, 22: 995–1012.

Bühlmann, P. and Künsch, H. R. (1999). Block Length Selection in the Bootstrap for Time Series. *Computational Statistics and Data Analysis*, 31: 295–310.

Bullmore, E., Long, C., Suckling, J., Fadili, J., Calvert, G., Zelaya, F., Carpenter, T. A., and Brammer, M. (2001). Colored Noise and Computational Inference in Neurophysiological (fMRI) Time Series Analysis: Resampling Methods in Time and Wavelet Domains. *Human Brain Mapping*, 12: 61–78.

Burke, M. D. (2000). Multivariate Tests-of-Fit and Uniform Confidence Bands using a Weighted Bootstrap. *Statistics and Probability Letters*, 46(1): 13–20.

Cabrera, J. and Meer, P. (1996). Unbiased Estimation of Ellipses by Bootstrapping. *IEEE Transactions on Pattern Analysis and Machine Intelligence*, 18(7): 752–756.

Chatterjee, S. (1986). Bootstrapping ARMA Models: Some Simulations. *IEEE Transactions on Systems, Man, and Cybernetics*, 16(2): 294–299.

Chen, D.-R., Kuo, W.-J., Chang, R.-F., Moon, W. K., and Lee, C. C. (2002). Use of the Bootstrap Technique with Small Training Sets for Computer-Aided Diagnosis in Breast Ultrasound. *Ultrasound in Medicine and Biology*, 28(7): 897–902.

Chen, H. and Romano, J. P. (1999). Bootstrap-Assisted Goodness-of-Fit Tests in the Frequency Domain. *Journal of Time Series Analysis*, 20(6): 619–654.

Chernick, M. R. (1999). *Bootstrap Methods*. John Wiley and Sons.

Chung, H.-C. and Han, C.-P. (2000). Discriminant Analysis when a Block of Observations is Missing. *Annals of the Institute of Statistical Mathematics*, 52(3): 544–556.

Coakley, K. J. (1996). Bootstrap Method for Nonlinear Filtering of EM-ML Reconstructions of PET Images. *International Journal of Imaging Systems and Technology*, 7(1): 54–61.

Conte, E., Di Bisceglie, M., Longo, M., and Lops, M. (1995). Canonical Detection in Spherically Invariant Noise. *IEEE Transactions on Communications*, 43: 347–353.

Conte, E., Lops, M., and Ricci, G. (1996). Adaptive Matched Filter Detection in Spherically Invariant Noise. *IEEE Signal Processing Letters*, 3(8): 248–250.

Cramér, H. (1999). *Mathematical Methods of Statistics*. Princeton University Press, Princeton, N.J.

Daniels, D. J., Gunton, D. J., and Scott, H. F. (1988). Introduction to Subsurface Radar. *Proceedings of IEE*, 135(F): 278–317.

Davison, A. C. and Hinkley, D. V. (1997). *Bootstrap Methods and their Application*. Cambridge Series in Statistical and Probabilistic Mathematics, No. 1, Cambridge University Press, Cambridge, NY.

Davison, A. C., Hinkley, D. V., and Schechtman, E. (1986). Efficient Bootstrap Simulation. *Biometrika*, 73: 555–566.

De Angelis, D. and Young, G. A. (1992). Smoothing the Bootstrap. *International Statistical Review*, 60: 45–56.

Dejian, L. and Guanrong, C. (1995). Computing the Distribution of the Lyapunov Exponent from Time Series: The One-Dimensional Case Study. *International Journal of Bifurcation and Chaos in Applied Sciences and Engineering*, 5: 1721–1726.

DiFranco, J. V. and Rubin, W. L. (1980). *Radar Detection*. Artech House, Dedham, Mass.

Djurić, P. M. (1997). Using the Bootstrap to Select Models. In *Proceedings of the IEEE International Conference on Acoustics, Speech and Signal Processing, ICASSP-97*, 5: 3729–3732, Munich, Germany.

Do, K. A. and Hall, P. (1991). On Importance Resampling for the Bootstrap. *Biometrika*, 78: 161–167.

Downton, M. W. and Katz, R. W. (1993). A Test for Inhomogeneous Variance in Time-Averaged Temperature Data. *Journal of Climate*, 6(12): 2448–2464.

Dupret, G. and Koda, M. (2001). Bootstrap Resampling for Unbalanced Data in Supervised Learning. *European Journal of Operational Research*, 134(1): 141–156.

Dzhaparidze, K. O. and Yaglom, A. M. (1983). Spectrum Parameter Estimation in Time Series Analysis. In *Developments in Statistics*. Krishnaiah, P. R., Editor, Volume 4, Academic Press, New York.

Efron, B. (1979a). Bootstrap Methods. Another Look at the Jackknife. *The Annals of Statistics*, 7: 1–26.

  (1979b). Computers and the Theory of Statistics: Thinking the Unthinkable. *SIAM Review*, 4: 460–480.

  (1981). Nonparametric Standard Errors and Confidence Intervals (with Discussion). *The Canadian Journal of Statistics*, 9: 1–26.

  (1982). The Jackknife, the Bootstrap and Other Resampling Plans. CBMS Monograph 38, Society for Industrial and Applied Mathematics, Philadelphia.

  (1987). Better bootstrap confidence interval. *Journal of the American Statistical Association*, 82: 171–185.

  (1990). More efficient bootstrap computations. *Journal of the American Statistical Association*, 85: 79–89.

  (2002). The Bootstrap and Modern Statistics. In *Statistics in the 21st century*, Raftery, A. E., Tanned, M., Wells M., Editors, 326–332, Chapman and Hall/CRC, Boca Raton, Fla.

Efron, B. and Gong, G. (1983). A Leisurely Look at the Bootstrap, the Jackknife, and Cross-Validation. *The American Statistician*, 37: 36–48.

Efron, B. and Tibshirani, R. (1993). *An Introduction to the Bootstrap*. Chapman and Hall, New York.

Ferguson, B. C. (1992). A Ground Based Narrow-Band Passive Acoustic Technique for Estimating the Altitude and Speed of a Propeller Driven Aircraft. *Journal of the Acoustic Society of America*, 92(3): 1403–1407.

Ferguson, B. C. and Quinn, B. G. (1994). Application of the Short-Time Fourier Transform and the Wigner-Ville Distribution to the Acoustic Localization of Aircraft. *Journal of the Acoustic Society of America*, 96(2): 821–827.

Field, C. and Ronchetti, E. (1990). *Small Sample Asymptotics*. Institute of Mathematical Statistics, Hayward, CA.

Fisher, N. I. and Hall, P. (1989). Bootstrap Confidence Regions for Directional Data. *Journal of the American Statistical Society*, 84: 996–1002.

(1990). New Statistical Methods for Directional Data – I. Bootstrap Comparison of Mean Directions and the Fold Test in Palaeomagnetism. *Geophysics Journal International*, 101: 305–313.

(1991). General Statistical Test for the Effect of Folding. *Geophysics Journal International*, 105: 419–427.

Fisher, R. A. (1970). *Statistical Methods for Research Workers*. 14th Edition. Oliver and Boyd, Edinburgh.

Fishler, E. and Messer, H. (1999). Order Statistics Approach for Determining the Number of Sources Using an Array of Sensors. *IEEE Signal Processing Letters*, 6(7): 179–182.

Franke, J. and Härdle, W. (1992). On Bootstrapping Kernel Estimates. *The Annals of Statistics*, 20: 121–145.

Franke, J. and Neumann, M. H. (2000). Bootstrapping Neural Networks. *Neural Computation*, 12(8): 1929–1949.

Freedman, D. A. (1981). Bootstrapping Regression Models. *The Annals of Statistics*, 9: 1218–1228.

Friedlander, B. and Porat, B. (1995). The Sensitivity of High-Order Parameter Estimation to Additive Gaussian Noise. In *Higher Order Statistical Signal Processing*, Boashash, B., Powers, E. J., and Zoubir, A. M., Editors, Longman Cheshire, 89–111.

Fujikoshi, Y. (1980). Asymptotic Expansions for the Distributions of the Sample Roots under Nonnormality. *Biometrika*, 67(1): 45–51.

Gasser, T. (1975). Goodness-of-Fit Tests for Correlated Data. *Biometrika*, 62: 563–570.

Gibson, J. D. and Melsa, J. L. (1996). *Introduction to Nonparametric Detection with Applications*. IEEE Press, New York.

Gine, E. and Zinn, J. (1989). Necessary Conditions for the Bootstrap of the Mean. *The Annals of Statistics*, 17(2): 684–691.

Goodman, N. R. (1963). Statistical Analysis based on a Certain Multivariate Complex Gaussian Distribution. *Annals of Mathematical Statistics*, 34: 152–171.

Guera, R., Polansky, A. M., and Schucany, W. R. (1997). Smoothed Bootstrap Confidence Intervals with Discrete Data. *Computational Statistics and Data Analysis*, 26(2): 163–176.

Hahn, G. J. and Shapiro, S. S. (1967). *Statistical Models in Engineering*. J. Wiley & Sons,

Hall, P. (1988). Theoretical Comparison of Bootstrap Confidence Intervals (with Discussion). *The Annals of Statistics*, 16: 927–985.

(1989a). Antithetic Resampling for Bootstrap. *Biometrika*, 76: 713–724.

(1989b). Bootstrap Methods for Constructing Confidence Regions for Hands. *Communications in Statistics. Stochastic Models*, 5(4): 555–562.

(1992). *The Bootstrap and Edgeworth Expansion*. Springer-Verlag New York, Inc., New York.

(2001). Biometrika Centenary: Nonparametrics. *Biometrika*, 88(1): 143–165.

Hall, P. and Jing, B. Y. (1994). On Sample Re-Use Methods for Dependent Data. *Statistic Report No. SR8-94*, Centre for Mathematics and its Applications, ANU, Canberra.

Hall, P. and Maesono, Y. (2000). A Weighted Bootstrap Approach to Bootstrap Iteration. *Journal of the Royal Statistical Society, Series B*, 62(1): 137–144.

Hall, P. and Titterington, D. M. (1989). The Effect of Simulation Order on Level Accuracy and Power of Monte Carlo Tests. *Journal of the Royal Statistical Society, Series B*, 51: 459–467.

Hall, P. and Wilson, S. R. (1991). Two Guidelines for Bootstrap Hypothesis Testing. *Biometrics*, 47: 757–762.

Hanna, S. R. (1989). Confidence Limits for Air Quality Model Evaluations, as Estimated by Bootstrap and Jackknife Resampling Methods. *Atmospheric Environment*, 23: 1385–1398.

Hannan, E. J. and Quinn, B. G. (1979). The Determination of the Order of an Autoregression. *Journal of the Royal Statistical Society, Series B*, 41: 190–195.

Hastie, T. and Tibshirani, R. (1990). *Generalized Additive Models*. Chapman & Hall, London, New York.

Haynor, D. R. and Woods, S. D. (1989). Resampling Estimates of Precision in Emission Tomography. *IEEE Transactions on Medical Imaging*, 8: 337–343.

Hens, N., Aerts, M., Claeskens, G., and Molenberghs, G. (2001). Multiple Nonparametric Bootstrap Imputation. In *Proceedings of the 16th International Workshop on Statistical Modelling*, 219–225.

Hero, A. O., Fessler, J. A., and Usman M. (1996). Exploring Estimator Bias-Variance Tradeoffs using the Uniform CR Bound. *IEEE Transactions on Signal Processing*, 44(8): 2026–2042.

Hewer, G. A., Kuo, W., and Peterson, L. A. (1996). Multiresolution Detection of Small Objects using Bootstrap Methods and Wavelets. In Drummond, O. E., Editor, *Signal and Data Processing of Small Targets*, Proceedings of SPIE, 2759: 2–13, San Diego.

Hinkley, D. V. (1988). Bootstrap Methods. *J. Royal Statistical Society, Series B*, 50(3): 321–337.

Hochberg, Y. and Tamhane, A. (1987). *Multiple Comparison Procedures*. John Wiley, 1987.

Holm, S. (1979). A Simple Sequentially Rejective Multiple Test Procedure. *Scandinavian Journal of Statistics* 6: 65–70.

Hu, F. and Hu, J. (2000). A note on Breakdown Theory for Bootstrap Methods. *Statistics and Probability Letters*, 50: 49–53.

Hurvich, C. M. and Tsai, C. L. (1989). Regression and Time Series Model Selection in Small Samples. *Biometrika*, 76: 297–307, 1989.

ICASSP-94 Special Session. The bootstrap and its applications. In *Proceedings of the IEEE International Conference on Acoustics, Speech and Signal*

﹁ *Processing, ICASSP-94*, 6: 65–79, Adelaide, Australia.

Iskander, D. R. (1998). On the Use of a General Amplitude pdf in Coherent Detectors of Signals in Spherically Invariant Interference. In *Proceedings of the IEEE International Conference on Acoustics, Speech and Signal Processing, ICASSP-98*, 6: 2145–2148.

Iskander, D. R., Arnold, M. J., and Zoubir, A. M. (1995). Testing Gaussianity Using Higher Order Statistics: The Small Sample Size Case. In *Proceedings of the IEEE Signal Processing/ATHOS Workshop on Higher-Order Statistics*, 191–195, Begur, Spain.

Iskander, D. R., Collins, M. J., Davis, B., and Carney, L. G. (2000a). Monochromatic Aberrations and Characteristics of Retinal Image Quality. *Clinical and Experimental Optometry*, 83(6): 315–322.

Iskander, D. R., Zoubir, A. M., and Chant I. (2000b). Time-Varying Spectrum Based Detection of Landmines Using Ground Penetrating Radar. In *Proceedings of Eighth International Conference on Ground Penetrating Radar, GPR 2000*, 65–68, Gold Coast, Australia.

Iskander, D. R., Collins, M. J., and Davis, B. (2001). Optimal Modeling of Corneal Surfaces with Zernike Polynomials. *IEEE Transactions on Biomedical Engineering*, 48(1): 87–95.

Iskander, D. R., Morelande, M. R., Collins, M. J., and Davis, B. (2002). Modeling of Corneal Surfaces with Radial Polynomials. *IEEE Transactions on Biomedical Engineering*, 49(4): 320–328.

Iskander, D. R., Morelande, M. R., Collins, M. J., and Buehren, T. (2004). A Refined Bootstrap Method for Estimating the Zernike Polynomial Model Order for Corneal Surfaces. *IEEE Transactions on Biomedical Engineering*, (in press).

Jaeckel, L. B. (1971). Some Flexible Estimates of Location. *Annals of Mathematical Statistics*, 42: 1540–1552.

Jain, A. K., Dubes, R. C., and Chen, C. C. (1987). Bootstrap Techniques for Error Estimation. *IEEE Transactions on Pattern Analysis and Machine Intelligence*, 9: 628–633.

Jakeman, E. and Pusey, P. N. (1976). A Model for Non-Rayleigh Sea Echo. *IEEE Transactions on Antennas and Propagation*, 24: 806–814.

James, M. (1960). The Distribution of the Latent Roots of the Covariance Matrix. *The Annals of Mathematical Statistics*, 31: 151–158, 1960.

Johnson, N. L. and Kotz, S. (1970). *Continuous Univariate Distributions*, Volume 2, Wiley, New York.

Jones, L. A. and Woodall, W. H. (1998). The Performance of Bootstrap Control Charts in Quality Control. *Journal of Quality Technology*, 30(4): 362–375.

Karlsson, S. and Löthgren, M. (2000). Computationally Efficient Double Bootstrap Variance Estimation. *Computational Statistics and Data Analysis*, 33: 237–247, 2000.

Kawano, H. and Higuchi, T. (1995). The Bootstrap Method in Space Physics: Error Estimation for the Minimum Variance Analysis. *Geophysical Research Letters*, 22(3): 307–310.

Kay, S. M. (1993). *Fundamentals of Statistical Signal Processing. Estimation Theory*, Volume I. Prentice Hall, Englewood Cliffs, New Jersey.

   (1998). *Fundamentals of Statistical Signal Processing. Detection Theory*, Volume II. Prentice Hall, Englewood Cliffs, New Jersey.

Kazakos, D. and Papantoni-Kazakos, P. (1990). *Detection and Estimation.*

Computer Science Press, New York.

Kendall, M. G. and Stuart, A. (1967). *The Advanced Theory of Statistics*. Charles Griffin & Company, London.

Khatri, C. G. (1965). Classical Statistical Analysis based on Certain Multivariate Complex Distributions. *Annals of Mathematical Statistics*. 36: 98–114, 1965.

Kijewski, T. and Kareem, A. (2002). On the Reliability of a Class of System Identification Techniques: Insights from Bootstrap Theory. *Structural Safety*, 24: 261–280.

Kim, J. K. (2002). A Note on Approximate Bayesian Bootstrap Imputation. *Biometrika*, 89(2): 470–477.

Kim, Y. and Singh, K. (1998). Sharpening Estimators Using Resampling. *Journal of Statistical Planning and Inference*, 66: 121–146.

Knuth, D. E. (1981). *The Art of Computer Programming*. Addison-Wesley, Reading, Mass.

König, D. and Böhme, J. F. (1996). Wigner-Ville Spectral Analysis of Automotive Signals Captured at Knock. *Applied Signal Processing*, 3: 54–64.

Kreiss, J. P. and Franke, J. (1992). Bootstrapping Stationary Autoregressive Moving Average Models. *Journal of Time Series Analysis*, 13: 297–319.

Krolik, J. L. (1994). Sensor Array Processing Performance Evaluation via Bootstrap Resampling. *Journal of the Acoustic Society of America*, 95(2): 798–804.

Krolik, J., Niezgoda, G., and Swingler, D. (1991). A Bootstrap Approach for Evaluating Source Localization Performance on Real Sensor Array Data. In *Proceedings of International Conference on Acoustics, Speech and Signal Processing, ICASSP-91*, 2: 1281–1284, Toronto, Canada.

Krzyscin, J. W. (1997). Detection of a Trend Superposed on a Serially Correlated Time Series. *Journal of Atmospheric and Solar Terrestrial Physics*, 59(1): 21–30.

Künsch, H. R. (1989). The Jackknife and the Bootstrap for General Stationary Observations. *The Annals of Statistics*, 17: 1217 1241.

Lanz, E., Maurer, H., and Green, A. G. (1998). Refraction Tomography Over a Buried Waste Disposal Site. *Geophysics*, 63(4): 1414–1433.

Lawley, D. (1956). Tests of Significance for the Latent Roots of Covariance and Correlation Matrices. *Biometrika*, 43: 128–136, 1956.

LeBaron, B. and Weigend, A. S. (1998). A Bootstrap Evaluation of the Effect of Data Splitting on Financial Time Series. *IEEE Transactions on Neural Networks*, 9(1): 213–220.

Léger, C. and Romano, J. P. (1990). Bootstrap Adaptive Estimation: The Trimmed Mean Example. *Canadian Journal of Statistics*, 18: 297–314.

Léger, C., Politis, D., and Romano, J. (1992). Bootstrap Technology and Applications. *Technometrics*, 34: 378–398.

Lehmann, E. L. (1991). *Testing Statistical Hypotheses*, Wadsworth and Brooks, Pacific Grove, Calif.

LePage, R. and Billard, L. (1992). Editors, *Exploring the Limits of Bootstrap*. John Wiley and Sons, New York.

Linville, C. D., Hobbs, B. F., and Vekatesh, B. N. (2001). Estimation of Error and Bias in Bayesian Monte Carlo Decision Analysis Using the Bootstrap. *Risk Analysis*, 21(1): 63–74.

Liu, R. Y. and Singh, K. (1992). Moving Blocks Jackknife and Bootstrap Capture Weak Dependence. In *Exploring the Limits of Bootstrap*. LePage, R. and

Billard, L., Editors, John Wiley, New York.

Ljung, L. (1987). *System Identification. Theory for the User.* Prentice-Hall, Englewood Cliffs, NJ.

Locascio, J. J., Jennings, P. J., Moore, C. I., and Corkin, S. (1997). Time Series Analysis in the Time Domain and Resampling Methods for Studies of Functional Magnetic Resonance Imaging. *Human Brain Mapping*, 5: 168–193.

Maitra, R. (1998). An Approximate Bootstrap Technique for Variance Estimation in Parametric Images. *Medical Image Analysis*, 2(4): 379–393.

Mallows, C. L. (1973). Some comments on C_p. *Technometrics*, 15: 661–675.

Malti, R., Ekongolo, S. B., and Ragot, J. (1998). Dynamic SISO and MISO System Approximations Based on Optimal Laguerre Models. *IEEE Transactions on Automatic Control*, 43(9): 1318–1323.

Mammen, E. (1992). When does Bootstrap work? Asymptotic Results and Simulations. *Lecture notes in statistics*, Volume 77, Springer, New York.

Mandell, R. B. (1996). A Guide to Videokeratography. *International Contact Lens Clinic*, 23(6): 205–228.

Manly, B. F. J. (1997). *Randomization, Bootstrap and Monte Carlo Methods in Biology.* Chapman and Hall/CRC.

Manoukian, E. B. (1986). *Modern Concepts and Theorems of Mathematical Statistics.* Springer-Verlag, New York.

Marple Jr, S. L. (1987). *Digital Spectral Analysis with Applications.* Prentice-Hall, Englewood Cliffs, NJ.

Mehrotra, K., Jackson, P., and Schick, A. (1991). On Choosing an Optimally Trimmed Mean. *Communications in Statistics. Simulation and Control*, 20: 73–80.

Middleton, M. (1999). Non-Gaussian Noise Models in Signal Processing for Telecommunications: New Methods and Results for Class A and Class B Noise Models. *IEEE Transactions on Information Theory*, 45(4): 1129–1149.

Miller, J. H. and Thomas, J. B. (1972). Detectors for Discrete-time Signals in Non-Gaussian Noise. *IEEE Transactions on Information Theory*, 18(2): 241–250.

Miller, R. G. (1974). The Jackknife - A Review. *Biometrika*, 61: 1–15.

Ming, C. L. and Dong, S. C. (1997). Bootstrap Prediction Intervals for the Birnbaum-Saunders Distribution. *Microelectronics and Reliability*, 37(8): 1213–1216.

Mudelsee, M. (2000). Ramp Function Regression: a Tool for Quantifying Climate Transitions. *Computer and Geosciences*, 26(3): 293–307.

Nagaoka, S. and Amai, O. (1990). A Method for Establishing a Separation in Air Traffic Control Using a Radar Estimation Accuracy of Close Approach Probability. *Journal of the Japan Institute of Navigation*, 82: 53–60.

(1991). Estimation Accuracy of Close Approach Probability for Establishing a Radar Separation Minimum. *Journal of Navigation*, 44: 110–121.

Neyman, J. and Pearson, E. S. (1928). On the Use and Interpretation of Certain Test Criteria for Purposes of Statistical Interference. *Biometrika*, 20: 175–240.

Noll, R. J. (1976). Zernike Polynomials and Atmospheric Turbulence. *Journal of the Optical Society of America*, 66(3): 207–211.

Ong, H.-T. (2000). *Bootstrap Methods for Signal Detection.* PhD Thesis, Curtin University of Technology, Australia.

Ong, H.-T. and Zoubir, A. M. (1999a). CFAR Detector for Target Returns with

Unknown Phase and Frequency in Unspecified Non-Gaussian Noise. In Kadar, I., Editor, *Signal Processing, Sensor Fusion, and Target Recognition VIII*, 3720: 483–492, San Diego. Proceedings of SPIE.

(1999b). Robust Signal Detection Using the Bootstrap. In *IEEE International Conference on Acoustics, Speech, and Signal Processing. Proceedings, ICASSP-99*, 3: 1197–1200, Phoenix, Arizona.

(2000a). The Bootstrap Matched Filter and its Accuracy. *IEEE Signal Processing Letters*, 7(1): 11–13.

(2000b). Bootstrap Methods for Adaptive Signal Detection. In *IEEE International Conference on Acoustics, Speech, and Signal Processing. Proceedings, ICASSP-2000*, 1: 57–60, Istanbul, Turkey.

(2003). Bootstrap-based Detection of Signals with Unknown Parameters in Unspecified Correlated Interference. *IEEE Transactions on Signal Processing*, 51(1): 135–141.

Papadopoulos, G., Edwards, P. J., and Murray, A. F. (2001). Confidence Estimation Methods for Neural Networks: a Practical Comparison. *IEEE Transactions on Neural Networks*, 12(6): 1278–1287.

Paparoditis, E. (1996a). Bootstrapping Autoregressive and Moving Average Parameter Estimates of Infinite Order Vector Autoregressive Processes. *Journal of Multivariate Analysis*, 57: 277–296.

(1996b). Bootstrapping Periodogram and Cross Periodogram Statistics of Vector Autoregressive Moving Average Models. *Statistics and Probability Letters*, 27: 385–391.

Paparoditis, E. and Politis, D. N. (1999). The Local Bootstrap for Periodogram Statistics. *Journal of Time Series Analysis*, 20(2): 193–222.

Park, D. and Willemain, T. R. (1999). The Threshold Bootstrap and Threshold Jackknife. *Computational Statistics and Data Analysis*, 31: 187–202.

Park, S. K. and Miller, K. W. (1988). Random Number Generators: Good ones are hard to find. *Communications of the A. C. M.*, 32: 1192–1201.

Peleg, S. and Friedlander, B. (1995). The Discrete Polynomial-Phase Transform. *IEEE Transactions on Signal Processing*, 43: 1901–1914.

Politis, D. N. (1998). Computer-Intensive Methods in Statistical Analysis. *IEEE Signal Processing Magazine*, 15: 39–55.

Politis, D. N. and Romano, J. P. (1992a). A General Resampling Scheme for Triangular Arrays of α-mixing Random Variables with Application to the Problem of Spectral Density Estimation. *The Annals of Statistics*, 20: 1985–2007.

(1992b). Circular Block-Resampling Procedure for Stationary Data. In *Exploring the Limits of Bootstrap*. LePage, R. and Billard, L., Editors, John Wiley, New York.

(1994). The Stationary Bootstrap. *Journal of the American Statistical Association*, 89(428): 1303–1313.

Politis, D. N., Romano, J. P., and Lai, T.-L. (1992). Bootstrap Confidence Bands for Spectra and Cross-Spectra. *IEEE Transactions on Signal Processing*, 40(5): 1206–1215.

Politis, D. N., Romano, J. P., and Wolf, M. (1999). *Subsampling*. Springer, New York.

Poor, H. V. (1986). Robustness in signal detection. In *Communications and Networks: A Survey of Recent Advances*, Blake, I. F. and Poor, H. V., Editors, 131–156, Springer-Verlag.

(1994). *An introduction to signal detection and estimation.* Springer-Verlag, New York.

Porat, B. (1994). *Digital Processing of Random Signals. Theory and Methods,* Englewood Cliffs, NJ.

Porat, B. and Friedlander, B. (1989). Performance Analysis of Parameter Estimation Algorithms Based on High-Order Moments. *International Journal on Adaptive Control and Signal Processing,* 3: 191–229.

Priestley, M. B. (1981). *Spectral Analysis and Time Series.* Academic Press, London.

Rao, C. R. and Wu, Y. (1989). A Strongly Consistent Procedure for Model Selection in a Regression Problem. *Biometrika,* 76: 469–474.

Rao, C. R., Pathak, P. K., and Koltchinskii, V. I. (1997). Bootstrap by Sequential Resampling. *Journal of Statistical Planning and Inference,* 64: 257–281, 1997.

Rao, J. S. (2000). Bootstrapping to Assess and Improve Atmospheric Prediction Models. *Data Mining and Knowledge Discovery,* 4(1): 29–41.

International Committee of the Red Cross, (1996). Anti-Personnel Landmines – Friend or Foe? March.

Reid, D. C. (1997). *Improved Aircraft Flight Parameter Estimation Based on Passive Acoustic Techniques using Time-Frequency Methods.* PhD Thesis, Queensland University of Technology, Australia.

Reid, D. C., Zoubir, A. M., and Boashash, B. (1996). The Bootstrap Applied to Passive Acoustic Aircraft Parameter Estimation. In *Proceedings of International Conference on Acoustics, Speech and Signal Processing, ICASSP-96,* 6: 3154–3157, Atlanta.

(1997). Aircraft Parameter Estimation based on Passive Acoustic Techniques Using the Polynomial Wigner-Ville Distribution. *Journal of the Acoustic Society of America,* 102(1): 207–223.

Rihaczek, A. W. (1985). *Principles of High-Resolution Radar.* Peninsula Publishing, Los Altos, CA.

Rissanen, J. (1983). A Universal Prior for Integers and Estimating by Minimum Description Length. *The Annals of Statistics,* 11: 416–431.

(1989). *Stochastic Complexity and Statistical Inquiry.* World Scientific, Singapore.

Robert, C. P. and Casella, G. (1999). *Monte Carlo Statistical Methods.* Springer, New York.

Romano, J. P. (1988). A Bootstrap Revival of some Nonparametric Distance Tests. *Journal of the American Statistical Association,* 83: 698–708.

Rosenblatt, M. (1985). *Stationary Sequences and Random Fields.* Birkhauser, Boston.

Ruly L.-U. C., Letaief, K. B., and Murch, R. D. (2001). MISO CDMA Transmission with Simplified Receiver for Wireless Communication Handsets. *IEEE Transactions on Communications,* 49(5): 888–898.

Sacchi, M. D. (1998). A Bootstrap Procedure for High-Resolution Velocity Analysis. *Geophysics,* 63(5): 1716–1725.

Saradhi, V. V. and Murty, M. N. (2001). Bootstrapping for Efficient Handwritten Digit Recognition. *Pattern Recognition,* 34: 1047–1056.

Scharf, L. L. (1991). *Statistical Signal Processing. Detection, Estimation, and Time Series Analysis.* Addison Wesley, Reading, Mass.

Schenker, N. (1985). Qualms About Bootstrap Confidence Intervals. *Journal of the American Statistical Association,* 80(390): 360–361.

Schwarz, G. (1978). Estimating the Dimensions of a Model. *The Annals of Statistics*, 6: 461–464.

Sekine, M. and Mao, Y. (1990). *Weibull Radar Clutter*. P. Peregrinus Ltd, London.

Seppala, T., Moskowitz, H., Plante, R., and Tang, J. (1995). Statistical Process Control via the Subgroup Bootstrap. *Journal of Quality Technology*, 27(2): 139–153.

Serfling, R. J. (1980). *Approximation Theorems of Mathematical Statistics*. Wiley, New York.

Shao, J. (1996). Bootstrap Model Selection. *Journal of the American Statistical Society*, 91: 655–665.

Shao, J. and Tu, D. (1995). *The Jackknife and Bootstrap*. Springer Verlag, New York.

Shao, Q. and Yu, H. (1993). Bootstrapping the Sample Means for Stationary Mixing Sequences. *Stochastic Process Applications*, 48: 175–190.

Shenton, L. R. and Bowman, K. O. (1977). *Maximum Likelihood Estimation in Small Samples*. Charles Griffin, London.

Shi, P. and Tsai, C. L. (1998). A note on the Unification of the Akaike Information Criterion. *Journal of the Royal Statistical Society, Series B*, 60: 551–558.

Shibata, R. (1984). *Approximate Efficiency of a Selection Procedure for the Number of Regression Variables*. Biometrika, 71: 43–49.

Shumway, R. H. (1983). Replicated Time-Series Regression: An Approach to Signal Estimation and Detection. *Handbook of Statistics III: Time Series in the Frequency Domain*, Brillinger, D. R. and Krishnaiah, P. R., Editors, 383–408, North-Holland, Amsterdam, New York.

Silverman, B. W. (1986). *Density Estimation for Statistics and Data Analysis*. Chapman and Hall, London.

Singh, K. (1981). On the Asymptotic Accuracy of Efron's Bootstrap. *The Annals of Statistics*, 9(6): 1187–1195.

Skolnik, M. I. (1990). *Radar Handbook*. McGraw Hill, New York.

Tauxe, L., Kylstra, N., and Constable, C. (1991). Bootstrap Statistics for Palaeomagnetic Data. *Journal of Geophysical Research*, 96: 11723–11740.

Thomas, J. B. (1970). Nonparametric detection. *Proceedings of IEEE*, 58(5): 623–631.

Thomson, J. and Chave, D. (1991). Jackknifed Error Estimates for Spectra, Coherence, and Transfer Functions. In *Advances in Spectrum Analysis and Array Processing*, Haykin, S., Editor, 1: 58–113. Prentice Hall.

Tibshirani, R. (1988). Variance Stabilization and the Bootstrap. *Biometrika*, 75: 433–444, 1988.

　　(1992). Comment on "Two Guidelines for Bootstrap Hypothesis Testing", by P. Hall and S. R. Wilson. *Biometrics*, 48: 969–970.

　　(1996). A Comparison of Some Error Estimates for Neural Network Models. *Neural Computation*, 8: 152–163.

Tjärnström, F. and Ljung, L. (2002). Using the Bootstrap to Estimate the Variance in the Case of Undermodeling. *IEEE Transactions on Automatic Control*, 47(2): 395–398.

Ulrych, T. J. and Sacchi, M. D. (1995). Sompi, Pisarenko and the Extended Information Criterion. *Geophysical Journal International*, 122(3): 719–724.

Van Trees, H. L. (2001a). *Detection, Estimation, and Modulation Theory. Part I*. John Wiley and Sons.

(2002a). *Detection, Estimation, and Modulation Theory. Part II. Nonlinear Modulation Theory.* John Wiley and Sons.

(2001b). *Detection, Estimation, and Modulation Theory. Part III. Radar-Sonar Signal Processing and Gaussian Signals in Noise.* John Wiley and Sons.

(2002b). *Detection, Estimation, and Modulation Theory. Part IV. Optimum Array Signal Processing.* John Wiley and Sons.

Verotta, D. (1998). Characterizing the Variability of System Kernel and Input Estimates. *Annals of Biomedical Engineering*, 26(5): 870–882.

Waternaux, C. (1976). Asymptotic distribution of the sample roots for a nonnormal population. *Biometrika*, 63(3): 639–645, 1976.

Wax, M. and Kailath, T. (1985). Detection of Signals by Information Theoretic Criteria. *IEEE Transactions on Acoustics, Spccch and Signal Processing*, 33: 387–392.

White, H. and Racine, J. (2001). Statistical Inference, the Bootstrap, and Neural-Network Modeling with Application to Foreign Exchange Rates. *IEEE Transactions on Neural Networks*, 12(4): 657–673.

Wilks, S. S. (1941). Determination of Sample Size for Setting Tolerance Limits. *Annals of Mathematical Statistics*, 12: 91–96.

Williams, D. and Johnson, D. (1990). Using the Sphericity Test for Source Detection with Narrow-Band Passive Arrays. *IEEE Transactions on Acoustics, Speech and Signal Processing*, 38(11): 2008–2014.

Xu, C. W. and Shiue, W. K. (1991). Parallel Bootstrap and Inference for Means. *Computational Statistics Quarterly*, 3: 233–239.

Young, G. A. (1994). Bootstrap: More than a Stab in the Dark? *Statistical Science*, 9: 382–415.

Zhang, Y., Hatzinakos, D., and Venetsanopoulos, A. N. (1993). Bootstrapping Techniques in the Estimation of Higher-Order Cumulants from Short Data Records. In *Proceedings of the IEEE International Conference on Acoustics, Speech and Signal Processing, ICASSP-93*, 6: 200–203, Minneapolis, MN, USA.

Zhao, L. C., Krishnaiah, P. R., and Bai, Z. D. (1986). On Detection of the Number of Signals in Presence of White Noise. *Journal of Multivariate Analysis*, 20(1): 1–25.

Zoubir, A. M. (1993). Bootstrap: Theory and Applications. *Advanced Signal Processing Algorithms Architectures and Implementations*, Luk, T., Editor, Proceedings of SPIE, 2027: 216–235.

(1994). Multiple Bootstrap Tests and their Application. *Proceedings of the IEEE International Conference on Acoustics, Speech and Signal Processing, ICASSP-94*, 6: 69–72, Adelaide, Australia.

(1999). The Bootstrap: A Powerful Tool for Statistical Signal Processing with Small Sample Sets. *ICASSP-99 Tutorial.* Available at: http://www.csp.curtin.edu.au/downloads/bootstrap_tutorial.html.

(2001). Signal Detection Using the Bootstrap. In *Defence Applications of Signal Processing*, Cochran, D., Moran, W., and White, L. B., Editors, Elsevier Science, B. V., 309–315.

Zoubir, A. M. and Boashash, B. (1998). The Bootstrap: Signal Processing Applications. *IEEE Signal Processing Magazine*, 15(1): 56–76.

Zoubir, A. M. and Böhme, J. F. (1995). Multiple Bootstrap Tests: An Application to Sensor Location. *IEEE Transactions on Signal Processing*, 43(6): 1386–1396.

Zoubir, A. M. and Iskander, D. R. (1999). Bootstrapping Bispectra: An Application to Testing for Departure from Gaussianity of Stationary Signals. *IEEE Transactions on Signal Processing*, 47(3): 880–884.

   (2000). Bootstrap Modeling of a Class of Nonstationary Signals. *IEEE Transactions on Signal Processing*, 48(2): 399–408.

Zoubir, A. M., Iskander, D. R., Ristic, B., and Boashash, B. (1994a). Bootstrapping Confidence Bands for the Instantaneous Frequency. In *Advanced Signal Processing Algorithms, Architectures and Implementations*, Luk, T., Editor, 2296: 176–190, San Diego. Proceedings of SPIE.

Zoubir, A. M., Ralston, J. C., and Iskander, D. R. (1997). Optimal Selection of Model Order for Nonlinear System Identification using the Bootstrap. In *Proceedings of the International Conference on Acoustics, Speech and Signal Processing 1997 (ICASSP-97)*, 5: 3945–3948, Munich, Germany.

Zoubir, A. M., Iskander, D. R., Chant, I., and Carevic, D. (1999). Detection of Landmines Using Ground Penetrating Radar. In *Detection and Remediation Technologies for Mines and Minelike Targets IV*, Dubey, A. C., Harvey, J. F., Broach, J. T., and Dugan, R. E., Editors, 3710: 1301–1312, San Diego. Proceedings of SPIE.

Zoubir, A. M., Chant, I. J., Brown, C. L., Barkat, B., and Abeynayake, C. (2002). Signal Processing Techniques for Landmine Detection Using Impulse Ground Penetrating Radar. *IEEE Sensors Journal* 2(1): 41–51.

Index

Printed in the United States
By Bookmasters